OCCUPY NATION

ALSO BY TODD GITLIN

Nonfiction

Uptown: Poor Whites in Chicago (with Nanci Hollander)

The Whole World Is Watching:
Mass Media in the Making and Unmaking of the New Left

Inside Prime Time

The Sixties: Years of Hope, Days of Rage

Watching Television (editor)

The Twilight of Common Dreams:
Why America Is Wracked by Culture Wars

Media Unlimited: How the Torrent of Images and Sounds
Overwhelms Our Lives

Letters to a Young Activist

The Intellectuals and the Flag

The Bulldozer and the Big Tent: Blind Republicans,
Lame Democrats, and the Recovery of American Ideals

The Chosen Peoples: America, Israel, and the Ordeals
of Divine Election (with Liel Leibovitz)

Fiction

The Murder of Albert Einstein

Sacrifice

Undying

Poetry

Campfires of the Resistance:
Poetry from the Movement (editor)

Busy Being Born

OCCUPY NATION

*The Roots, the Spirit,
and the Promise of*
OCCUPY WALL STREET

TODD GITLIN

Photos throughout by the author
Special photo supplement by Victoria Schultz

*it*books

AN IMPRINT OF HARPERCOLLINS *PUBLISHERS*

*it***books**

Photos throughout courtesy of the author. Special photo supplement, pp. 251–260, © Victoria Schultz.

HarperCollins books may be purchased for educational, business, or sales promotional use. For information please write: Special Markets Department, HarperCollins Publishers, 10 East 53rd Street, New York, NY 10022.

FIRST EDITION

Library of Congress Cataloging-in-Publication Data is available upon request.

ISBN 978-0-06-220092-1

12 13 14 15 16 DIX/RRD 10 9 8 7 6 5 4 3 2 1

To those who ignited the flame
and those who carry it on

CONTENTS

Part Three: The Promise 159

PREFACE

Ever since I started publishing books and lecturing about the movements of the sixties, movements that I was deeply involved in—antiwar, civil rights, the New Left—journalists have been asking me how the latest demonstration or uprising measures up to those of decades past. I would reel off similarities and differences, but in truth, the differences generally impressed me more. I sometimes thought journalists were hunting for similarities, in fact, either because of a certain nostalgia of their own—they were old enough to remember the uprisings fondly—or precisely because they had missed the fabled years and wondered if the strangeness they'd always heard about, possibly from their parents, might return.

I taught at universities, and wrote. I threw myself occasionally into insurgent campaigns—antinuclear efforts, divestment from South Africa—but, for better or worse, I never saw any prospect of the past recycling. Personally, this was all right with me, since I rather liked the post-sixties life I was living, was not much given to nostalgia, and did not believe I was getting any younger. I cautioned journalists that history didn't repeat itself, that the whole constellation of circumstances that clicked in the sixties had been unique. Some political efforts I found rather more exhilarating than others, but not once did I think that more great upheavals were impending. America was what it was.

Then, in recent years, especially when speaking abroad, I got used to another question: In the face of the disasters that have afflicted not only your country but the rest of the world, why are Americans so inert? In the spring of 2011, after the overthrow of Hosni Mubarak, I had the opportunity to lecture in Cairo about media and revolutions, and got to breathe the air of Tahrir Square after the tear gas had cleared (and before it returned). Then I gave a number of talks in Germany on the American news media and the recent, and ongoing, financial crisis, and the question of America's somnolence kept cropping up. In truth, the reasons that there was no wide-spread opposition on the left in the United States were fairly obvious. The economic climate was frigid. There was oligarchy, but not dictatorship. Students—the motors of change in the 1960s—were working, sometimes more than one job, while in school, then encumbered with debt that discouraged them from stepping out of line. There was not much cultural opposition because the entire consumer culture was, in its coy way, oppositional or at least, supposedly, alternative. Anyone could buy, or upload, or tattoo the accoutrements of rebellion. Everything was sayable, so nothing much mattered. Political rebellion became uncool. The wars of the 1980s and thereafter, wars that would once have provoked mass opposition, were short wars fought by professional soldiers. The one that inspired the grandest opposition, the Iraq expedition, sputtered out because the movement hadn't a prayer of influencing the hell-bent crusade of George W. Bush to shock and awe a recalcitrant world. The midcentury traditions had vanished. On the left, at least, sarcasm had replaced moral seriousness. And so on.

So, when I heard on the Internet grapevine, sometime during the summer of 2011, that a call had been issued to occupy Wall Street, I didn't take it too seriously. I'd been out of the country for months and so had missed New York's Bloombergville encampment in June and the limited fuss it caused. On a drive across the country in July, the only signs of political restiveness I saw were from Tea Party and militia enthusiasts. Back in New York, my contemporaries weren't talking excitedly about any forthcoming occupation of Wall Street. I rather assumed the call was another one of those puffs of hope and cheerleading for good causes that blow through cyberspace several times a day. I wondered whether it had been issued by one of those tiny left-wing sectarian groups that specialize in thick headlines that rub off on your hands. Whatever might materialize in lower Manhattan on September 17 would likely discredit itself without half trying. All the reasons why there hadn't been any such movement before would still apply.

In fact, when I started visiting Zuccotti Park, going to marches and meetings, interviewing occupiers, including people I'd been told were movers and shakers, I discovered that such skepticism and cynicism were normal reactions even among those who'd showed up first and stayed longest. This didn't look like any rally or march, though it included rallies and marches. There were few so-called name speakers. The dramaturgy was different—and weirder, and more original. Once you hung out in the movement—and hanging out was the main way you entered it—and opened your eyes, you found that this wild ensemble was not exactly what you had thought at first. It was at once more earnest, more energetic, more frivolous, more original, and more mysterious. Disparagement and incomprehension were not hard to derive from news media who were drawn to the most garish and photogenic sights in the encampment, the outré and the scruffy, with their scatter of protest slogans, their anger at capitalism, their general disrespect, their apparent disorder.

Impressed by the new, I set out to understand. What I found astounded me, not because it was wholly new under the sun (although, in sum, it *was* remarkably new), and not just because the Occupy people could make the case that they were part of a global upheaval (I had talked to revolutionaries in Cairo and so knew very well that Liberty Square was not another site of the Arab Spring, though the improbability of those heady upheavals made an American improbability seem less impossible), and not just because the Occupiers were amazingly intense, or because everything they did was, in my view, right, but in no small part because they were surprising, they made me laugh, they touched me. I was unprepared for their sheer sprawl and inventiveness. In rapid order—or disorder—they produced a social phenomenon that did not feel like a fad, because a fad is a single style and Occupy was all kinds of movements at once, some more visible and some less. And what impressed me the most, and drew me closer, was the eruption of intelligence I encountered there.

This book is an initial report on something very much in progress. In part I, I introduce some movers and shakers, trace the human, social, ideological roots of the movement, as best I understand them, and explore how they relate to the whole political-cultural ecology that includes them, including the larger organizations around them, the political parties, and Wall Street itself. In part II, I explore the movement's spirit, so unusual in the annals of social movements, yet not without precedent: its leaderlessness, its nonviolence, its rituals and obsessions, its divisions over conventional politics, over reform and revolution. In part III, I make some arguments about what seem to me the most promising directions, and worry about perils. I worry with this movement, not just about it.

As I write, many moving parts of the Occupy movement are in motion. Prediction is for fools and the jaded. But give credit where credit is due. We talk a lot about entrepreneurship in America, but the glories go to those who make capital grow. Occupy is a different kind of entrepreneurship, a creative and cooperative endeavor, and it profits America by making human vitality grow. Like all such undertakings, it is not guaranteed of success. But whatever becomes of this remarkable movement once the headlines yellow and Twitter trends move on to the next reality-show–wedding-divorce two-step, America has surely become more interesting—less predictable, more open, more vigorous, thrilling, boisterous, and collaborative all at once; which is an achievement to celebrate, and an astonishment.

Todd Gitlin
February 2, 2012
New York City

PART ONE

TWO ENERGY CENTERS

1. Pioneers

Few if any of the few dozen pioneers who unrolled their sleeping bags on the stony rectangle of Zuccotti Park on September 17, 2001, expected their insurgency to bloom so quickly into a movement so vast. They didn't dare. Daring had a way of turning treacherous. Not three years earlier, many of them dared think, or at least hope, that the election of Barack Obama was going to change the course of the nation. They had surged into his 2008 campaign feeling "the audacity of hope" but one cabinet appointment and policy shortfall after another left them disappointed, then demoralized. Others in Zuccotti Park, far fewer, handfuls of self-guided revolutionaries, dared believe in direct democracy as a guiding principle for a fundamentally revamped society, though they knew in their bones that such a society was not going to spring up in a month or a season. They might have adopted as camp epigram these words from science-fiction writer Robert Anton Wilson, picked up as an e-mail salutation by forty-seven-year-old filmmaker Michael Fix, who threw himself into Occupy Wall Street virtually full-time (and, after the eviction managed their office nearby): "You should view the world as a conspiracy run by a very closely knit group of nearly omnipotent people, and you should think of those people as yourself and your friends."

"Something has been opened up, a kind of space nobody knew existed," said Yotam Marom of Occupy, less than four months later. "Something just got kind of unclogged." What took root in Zuccotti Park and then sent out lateral shoots from there was, as Anthony Barnett would write, "the combination of hi-tech networking and no-tech gathering." The intricate human experience of face-to-face meeting—with responsibility shared and authority challenged—was galvanizing. There was a public place to go to, where attention could readily be paid, and individuals had faces and stories; there were electronic communiqués in real time and electronic summons for emergencies. These people were not demonstrating—that is, showing authorities that they wanted something in common—but creating a space where leaders and ideas could emerge. As days went by and they became used to inhabiting this space, they became a sort of new tribe.

As the weeks and months went by, the movement's movers and shakers were astounded and overjoyed at what they had wrought. They had, first of all, endured. They had withstood scorn, busts, billy clubs, pepper spray, and evictions—and grown. They set up functional working groups and decision-making structures that, however outlandish in the eyes of traditionalist outsiders, kept their spaces running decently, for the most part. They set up live stream channels for 24/7 video images, along with Facebook pages, Tumblrs, Twitter feeds, all manner of social linkages. The movement's live streaming was like "reality TV on steroids," somebody said. People in and around the movement (and who was exactly *in* the movement anyway?) started newspapers and theoretical journals. They lived pell-mell in the grip of what sociologist Barrie Thorne, writing about the sixties, once called *event time*, hurtling from action to action with high fervor and much jubilation. Some burned out, others flocked in, and still others, in widening circles, took off, felt inspired. Talk about audacity, talk about hope. Something was happening, never mind that quizzical and sardonic journalists were stumbling around like Mr. and Ms. Jones not knowing what it was. In the words of a hand-painted sign held aloft in an Occupy support crowd by an intensely serious middle-aged woman in lower Manhattan on October 5, 2011: THIS IS THE 1ST TIME I'VE FELT HOPEFUL IN A VERY LONG TIME.

The sort of sea changes in public conversation that took three years to develop during the long-gone sixties—about brutal war, unsatisfying affluence, debased politics, and the suppressed democratic promise—took three weeks in 2011. At warp speed, all kinds of people felt that they needed to have opinions about the movement, what it was doing and saying, and what it ought to do and say. This was especially true in Manhattan, where not only Occupy's own electronic but local news amplified the word, and people throughout the outer boroughs heard that something interesting was going on near Wall Street, though just what it was wasn't exactly clear. Widening circles of people showed up at Zuccotti Park, volunteering, debating, seeing for themselves. Hundreds across the country—and in other countries—planted encampments of their own. (In this book, I'll focus on New York's Occupy Wall Street, though with forays to other places.) On designated, coordinated days, much larger numbers marched, tens of thousands at a time. Unknown numbers of others were taking the movement to heart, taking it as a moral challenge and a personal problem. Should they apply for jobs in finance? What, if anything, should they do about the 2012 elections? This was the intense magic of a social movement: not that people

talked only about the movement as something outside themselves, something that should think X or do Y or stop doing Z, but that they took to heart the moral challenge, What *will* you do?

However, before there was Occupy Wall Street, there was Wall Street, that vortex of human calculation and passion, portal to a vast network of connected minds and impulses, where vast fortunes are made by insiders who master the game of heads-we-win-tails-we-win-too.

Wall Street is symbolic, a whorl of opportunity-making and opportunity-breaking where anything that can be marketed is marketed. Through Wall Street and its opposite numbers in London, Hong Kong, and elsewhere, capital—that potent and mysterious intangible—circulates and cascades, ever in pursuit of the highest returns. Capital has no fixed address. High-flying traders can conduct their high-speed buying and selling and bundling from almost any node in a virtually seamless planetary web. Hunches, calculations, loans, debts, and accounting, creative and otherwise, are as borderless as financial crises, and for the same reasons. Yet still, humanity being a species that craves face-to-face company, at lunch and in bars and via water-cooler chat, a lion's share of the core decisions that entail the fates of nations are launched from a compact terrestrial neighborhood. The human passions of greed and fear cluster in lower Manhattan's ornamented stone high-rises and boxy glass-and-steel edifices, jammed up against one another, where the prowess of hotshots is tested and good fortune rewarded with the proceeds from other people's money. Here, in the great investment banks, the graduates of Ivy League colleges (among others) underwrite securities, arrange mergers and acquisitions, devise extraordinary varieties of tradable paper, and invest in every quantifiable phenomenon under the sun, all the better to finance the chiefs' Porsches and Picassos and the hulking beach houses in the Hamptons and the private jets to overfly ground traffic on the way there and the paneled, extravagant yachts in which they float free of national boundaries. The proprietors make extravagant use of the ever-replenished ingenuity of thousands of Ph.D.'s in mathematics and physics, who work for them designing financial *products* (as if they were tangible things), *securitizing* (rendering marketable), and *managing* (a word designed to soothe nerves if ever there were one) *risk* (a word connoting the sort of danger that can be managed). But, however rectilinear the high-rise incarnations of calculation, however ornate the physical facades, however nicely carpeted the suites and well-appointed the conference rooms, Wall Street has always been a feral place: a scramble of fortune-hunters sometimes partnering to help each other forward and sometimes betraying each other as they scramble toward pinnacles of wealth, pinnacles that recede, somehow, the higher they climb. For there is always somebody on the wrong end of a winning deal, and there is always someone who possesses more than you do.

For the movers of money and the makers of mega-deals who channel their animal spirits into elaborate games that they play with other people's money, shunting it around the world at lightning speed in pursuit of the main chance, collecting huge sums in packages of compensation and fees and options and parachutes by merging, shuffling, enlarging, and breaking up companies the way some people merge lanes, shuffle cards, and break promises, and also packaging combinations of risky securities, sharing the risk by opening up lines of contagion, hooking banks in Reykjavik to bad loans in Rapid City, the value of these securities being predicated on the value of other risky securities whose risks were unknown to, or misunderstood by, their buyers if not their sellers. For these financial corporations, known collectively in the vernacular as *Wall Street*, the three decades since 1980 were high-octane, high-testosterone, go-go years of fortune-making and out-contracting, megamansion-erecting and name-engraving, attention-getting philanthropy, meteoric rises and high flights, succeeded by home and business bankruptcies in the millions and rescues and bailouts in the billions, punctuated by occasional crashes and burns. All the gaudy fortune hunting took place under the half-horrified, half-envious gaze of the rest of America—the supermajorities who call themselves middle class—while most young people looked forward to lives spent stringing together part-time jobs when they couldn't get full-time jobs, and living in much more cramped style than they grew up in, and returning to nest with their parents, and scrambling to pay back their college loans, and, if they came from poverty in the first place, looking forward to the likelihood of more of the same.

Here, in lower Manhattan, beats the heart of America's business civilization. Through a circulatory system known as *investment*, wealth never rests, and as it moves it makes things happen in the material world, distributing enchantments and losses, performing over and over again capitalism's historically unprecedented ceremonies of magic. Wealth and its promises flow into Wall Street in order to course again outward, then inward again, endlessly flowing, and along the way, building houses, heating them, furnishing them, wiring them, equipping them, sending in cement mixers and moving trucks, opening factories on distant continents, closing them, chopping down forests, drilling oil, mining coal, building wind turbines and solar panels, offering myriad opportunities to generate yet more capital to make more things happen.

This whirling center of passions and calculations and deals is not confined to a few hundred acres in lower Manhattan (with neighboring precincts in New Jersey and Connecticut)—or to the webs that tie it to the competing centers in London, Hong Kong, and elsewhere. It is tethered to Washington. For the past three decades, those banks and insurance companies and financial divisions of other corporations were liberated to conduct their business more or less as they pleased because government regulations, put in place during the Great Depression in order to protect the public, were lifted. They were lifted at the behest of financial industry lobbyists and campaign contributors, and their economist collaborators who swept aside decades of obstacles in the belief that financial markets, like others, fundamentally regulated themselves, and that it would be integral to the self-regulatory process if investment banks, commercial banks, brokerages, and insurance companies were permitted to merge, so that savings and investment could be conducted under the same roof, an efficiency for financial capital that had been banned by the Glass-Steagall Act of 1933, one of the regulatory pillars of the New Deal. Wall Street has long been crash prone. Watered stock, unregulated investment pools, and easy money in margin loans (the equivalent of subprime mortgages) brought it to its knees in the twenties, then the conglomerate craze in the sixties. During the deregulatory years, which began in the late seventies, when Jimmy Carter was president, but accelerated at turbo speed during Ronald Reagan's terms, as inventive financiers like Michael Milken and arbitrageurs like Ivan Boesky made fortunes with leveraged buyouts, nifty little "products" nicknamed junk bonds, and trading in inside information, Wall Street and Washington became the systole and diastole of America's (and therefore much of the world's) political economy. A system evolved in which the top financiers administered to themselves the rewards of self-dealing, squeezed through revolving doors, practiced deregulation and administrative collusion, organized themselves into combinations in the name of competition, all of this cheered on or at least tolerated by a larger public panicky about falling behind and convinced, more or less, that its own interests would be served too if capital were unleashed. Wall Street became, and, despite the economic crisis that some call the Great Recession, continues to be, the place where the action is: the rush, the buzz, the allure, the electricity. Capital might be an abstraction—no one has ever laid eyes on one fleck of it—but it was a potent one, for on its expectations contracts could be let, debts repaid, risks taken, vast organizations stood up and heaved into motion. For the mass murderers of al-Qaeda, Wall Street's most conspicuous towers had been prime targets for their cinematic materializations of grandeur. Then, for a decade and counting, the missing World Trade Center, and the construction zones that loomed there, slowly filling up the spaces formed by the world's tallest absence, would become the area's chief tourist attractions. Wall Street was the canyon of dreams.

Then, anticipated by no one, and yet with astonishing speed, as if bursting through the crust of a volcano long thought to be extinct, Occupy Wall Street erupted. It turned out there was another dream, this one circulating on Twitter: "Dear Americans, this July 4th, dream of insurrection against corporate rule," with the hashtag #occupywallstreet.

Within weeks the upsurge took on the feel of a popular movement, with its flare-ups of solidarity and blazes of sudden commitment, its improvisations and personal attachments, its incandescent compound of indignation, joy, outrage, hope, ingenuity, and resolve, its spikes of passion and wild ideas. As a moment in time passed into a movement in history, it astounded everyone, not least its participants, many of whom had long pined for a world-changing social movement in the interest of equity, some of whom had experienced such things themselves, or hoped against hope that they might do so someday, but had not dared think that a major eruption was possible now, never imagined how quickly it might be possible for a movement to take off, turning the homely verb *occupy* into a rallying cry and making "We are the 99 percent!" a household phrase. They were, as one of them put it, rebooting history.

However, nothing comes out of nowhere. There are a few origin stories that converge, remarkably, in that zeitgeisty way in which people who don't know one another sometimes get more or less the same idea at the same time. In

February 2010, thirty-seven-year-old independent journalist David DeGraw posted on his own website a call for a 99 percent movement. "It blew up virally," he says, and got picked up by Alternet, a major left-wing alternative news service. In January 2011, his site got knocked offline by hacker attacks—emanating from someplace unknown, he says—and since his service provider along with hundreds of other sites were also incapacitated, the provider declined to host his site any longer unless he graduated to a more expensive arrangement. When DeGraw put out a call for help, the network Anonymous came to his rescue and set up a new site for him. In March 2011, DeGraw and an Anonymous subgroup, operating together under the name A99, brazenly called for an Operation Empire State Rebellion on Flag Day, June 14. They would organize bank protests and close accounts.

Meanwhile, New York City anti-austerity activists had been building up a critical mass since the spring. On May 12, several thousand marched around Wall Street, summoned by an online call from a coalition of small left-wing groups calling itself New Yorkers Against Budget Cuts (NYABC). It began:

> We hear it every day: There's no more money. No money to keep our senior centers open, pay our teachers, serve the homeless or help college students graduate. There's no revenue to pay for vital services or invest in creating jobs. Well, people across this country have had enough. There is no revenue crisis; there is an inequality crisis.

Independently, on June 9, in British Columbia, the editor of the anticorporate magazine *Adbusters,* collaborating with a colleague in Berkeley, fixed upon September 17, 2011, as a good day to Occupy Wall Street, specifying "BRING TENT." As was their insouciant wont—for twenty-two years, *Adbusters* had deployed design elegancies to campaign against not only advertising but against a society that turned advertising into its definitive art form—they Photoshopped a picture of a ballerina gorgeously poised on the back of the famous bronze statue of a hard-charging bull, that proud and ferocious shrine to Wall Street's belief in itself. They launched their call where calls are launched nowadays: on the Internet.

On June 14, without any known connection to A99 or the *Adbusters* call, NYABC—with the help of city employee unions and a group called Picture the Homeless— summoned about a hundred young activists and city workers to camp across the street from City Hall in the rain, protesting Mayor Michael Bloomberg's budget cuts (including the closing of twenty fire stations) and proposed layoffs of teachers and social service workers. The unions pulled out after a day, not wishing to contend with the law, and Picture the Homeless left after a week. The activists called their encampment Bloombergville, after Depression-era Hoovervilles, and claimed inspiration from the occupations of public space in Madison, Wisconsin; in Tunisia and Egypt, and most recently, Madrid. After two weeks, a remnant group of thirteen was arrested trying to blockade City Council members who were preparing to vote on budget cuts. While Bloombergville lasted, it governed itself by a direct-democratic general assembly, meeting twice daily.

Meanwhile, the night of June 13, A99 had declared that Operation Empire State Rebellion would occupy Zuccotti Park, a concrete rectangle taking up three-quarters of an acre located just north of Wall Street, and reconstructed after the attacks of September 11, 2001, to include built-in low stone benches and tables, and a soaring, abstract bright-red Mark di Suvero sculpture, called *Joie de Vivre*, but generally known as the red sculpture. They would launch from there a nonviolent action of indefinite duration, with four demands:

- End the campaign finance and lobbying racket
- Break up the Fed & Too Big to Fail banks
- Enforce RICO laws against organized criminal class
- Order Ben Bernanke to step down

Many grouplets call for many grand changes at many times and places. Came June 14, a grand total of sixteen people showed up at Zuccotti Park, formerly known as Liberty Plaza Park. DeGraw was one of four who were prepared to camp out. They had a tent and some folding chairs. The A99 group decided to try returning to Zuccotti Park on September 10. Some went off to join Bloombergville. When they found out about the *Adbusters* call for September 17, they decided to consolidate on that date.

Meetings ensued, leading to more meetings—among

them, an August 2 gathering on Bowling Green, and several five-hour-long general assemblies in Tompkins Square Park. Should they make demands? *Adbusters* said no. Those who gathered on Bowling Green had no idea what to expect. About the impending September 17 encampment, committed activists were "jaded and condescending," says Pablo Benson, a twenty-nine-year-old sociology instructor from Puerto Rico with an MA from the New School. Some of his friends scoffed, "flipped it off." He didn't care. Nor did he care that the media didn't care. He was one of those radical homesteaders later described by journalist Will Bunch: "people with quiet fantasies of a revolution and a communal lifestyle buried deep in their souls, stunned by the rapture of learning there were others who felt the same way, a discovery that only happened because of the rise of Internet social networking."

One organizer, Isham Christie, a Choctaw from Oklahoma with tight black curls and a trim beard, formerly cofounder of a revived Students for a Democratic Society at the University of North Dakota, lately arrived in New York as a graduate student and union organizer, inspired by the uprising in Egypt, wrote later: "Serious doubts plagued my mind . . . from the very beginning. Will people show up? Will Sept. 17 . . . be no more than a fight with cops? Will we be strong enough to actually take a space?" He "almost left the movement a couple of times." He took heart from reading Nietzsche on the subject of cheerfulness.

On the afternoon of September 17, Benson and Christie were two of several hundred who rallied across the street from the southern end of seventeenth-century Bowling Green, in front of the Museum of the American Indian. The fenced-in green is overstuffed with historical prefiguration. At its northern end stands the bronze statue of a bull in midcharge, icon of Wall Street rambunctiousness. The museum building also houses the US Bankruptcy Court for the Southern District of New York, something that the demonstrators might not have known, just as they were also most likely unaware that, in 1770, a two-ton equestrian statue of King George III was erected on the green, where it stood until five days after the Declaration of Independence was signed on July 4, 1776, whereupon a band of Sons of Liberty toppled it.

Now, on September 17, veterans of May 12 and Bloombergville were at Bowling Green aiming to topple other royalty. Most were surely aware of, and inspired by (that is, breathing in), recent precedents—the occupations of Madison, Wisconsin, back in December, of Cairo's Tahrir Square in January, and of Madrid's Puerta del Sol, all directed against bankrupt political classes. Most shared a sense of occasion, theater, and disbelief. A group of students drove in from Oberlin College in Ohio, others came from Tennessee. A map circulated, listing three possible destinations. Listed first was the superblock known as One Chase Manhattan Plaza, which is privately owned, housing as it does the headquarters of the investment bank now known as JPMorgan Chase, the wealthiest bank—indeed, largest company—in the world; it holds, at this writing, $2 trillion in assets. However, the whole square block had been sealed off the night before by a steel fence. Presumably, there were police on the OWS mailing list.

Plan B was to head for Zuccotti Park, which enjoys New York City's curious designation, a Privately Owned Public Space, meaning that a real estate company received permission to build higher and more densely than usual in exchange for preserving a diminutive open space close by, open to the public in perpetuity. Thus Zuccotti Park, named for the US chairman of the Canadian-American owners, must, by law, be kept open, not subject to curfews, twenty-four hours per day. It was unfenced.

Here, a few dozen marchers settled down for the night and declared that they stood for the 99 percent against the 1 percent. A likely inspiration was the Nobel-winning economist Joseph E. Stiglitz's much-noticed article, "Of the 1%, by the 1%, for the 1%," published in the May issue of *Vanity Fair*. Couching the issue this way was a stroke of messaging genius, since it turned the tables on right-wingers who insisted that any campaign for economic justice and progressive taxation amounted to class warfare. If the 1 percent were responsible for rampant inequality, then the status quo was not warfare at all, but a rout. The message left lots of questions and ambiguities, of course. Who were the 1 percent anyway? By one calculation, 31 percent of them were executives of nonfinancial corporations, and 14 percent of financial institutions; together, these business executives owned half of all stocks and mutual funds. Were the 1 percent rotten individuals? Was it right to hate them? Could a 1 percenter redeem him- or herself? Did Warren Buffett turn into a righteous 1 percenter when he said publicly that it was unjust that his secretary was taxed at a higher rate than he was? And then, too: Were doctors (16 percent) or artists and celebrities (2 percent) who belonged to the 1 percent as damnable as investment bankers or oil company CEOs? More profoundly, did the moral flaw lie in the fact that some individuals had acquired vast wealth, or that they had done so while obeying the dictates of perverse incentives that rewarded them for taking blind risks of driving the global economy into the ground? And even more pointedly, was good old-fashioned greed at fault, or was it the way in which this staple of deadly sins was encouraged to flourish and outdo itself in recent decades? Or was the real economic problem a sickness of institutions, the grotesque fact that society's central decisions for allocating resources were made by the onrushing flow of capital that was bound to leave destruction, creative or not, in its wake; that these decisions were made by the dictates of a system that was indifferent to where the chips fell and on whom, a system that would corrupt saints?

The night was cold. Some people slept, but Benson didn't, although he had brought along a sleeping bag, New York not permitting the erection of tents in public. "The drum circle didn't help," he told me.

Who were these scourges of the wealthy and powerful who rallied in Bowling Green and made their way to Zuccotti Park with much brio and without clear ex-pectations? Even sympathetic observers struggled to get their minds around this blur of a phenomenon that so conspicuously failed to match their ideas of what a pro-test was supposed to look like and sound like. A visiting friend from Paris, veteran of decades of left-wing activ-ity, looked at the massive demonstration of November 17 and said: "I have never seen a political movement that is so apolitical." Right-wing journalists saw "goddamned hippies." Demographically speaking, there were no re-liable statistics, but by inspection the Occupiers were mostly young and mostly unemployed and therefore had time for encampments as well as a hunger for con-nection. One of them, Pablo Benson, twenty-nine, got specific. The key demographic that first night in Zuc-cotti Park, he thought, consisted of recent college gradu-ates encumbered with huge loans. "You?" I asked. "Of course!" he shot back with a grin. Some in the original crowd belonged to the free-floating population of radical New Yorkers, many living in lower-rent areas of Brook-lyn, veterans of one or another demonstration, some of them buddies, some strangers. They were mostly white: It might well have been a coincidence that Benson and Isham Christie were both sons of colonized populations. More than a few were artists of a downtown or Williams-burg disposition. They were the kind of people who, had they been in the audience when Bruce Springsteen shouted out his wake-up live-concert line, *Is there any-body alive out there?* over the years, would have thought,

Yeah, good question—and at some point answered: *The rest of the country may be dead, but I'm alive, and I know some other people.*

In later weeks, Occupiers would be heard to complain that they were met at first with a media blackout. According to Will Bunch, a top official at National Public Radio had justified an initial decision not to cover the protests in Lower Manhattan, insisting that there had not "been 'large numbers' of protesters; there have not been any 'prominent people' involved; there has not been any great disruption; (and) the protesters have failed to articulate a clear point or aim."

But, as Bill Dobbs, one of the Occupy PR Working Group and an experienced New York activist and PR pro, told me later: "There was never a blackout. You can't say that that the Old Gray Lady"—the *New York Times*—"didn't notice." (In fact, the *Times* ran a substantial report and photo on p. A22 of the Sunday paper—more than it typically gave demonstrations of equivalent size.) The *Daily News,* the *New York Post,* and local television also covered the Zuccotti occupation. Even though A99 had boasted that twenty thousand people would show up, and the actual turnout was a small fraction of that, Dobbs says, "You'd have to say that this demonstration was well covered. Especially when you realize that the people who put this demo together were either neutral or hostile to the mainstream media. They were not paying much attention to them at all. Ambivalence about mainstream media is part of this story." It wasn't until day five or six that Dobbs made a cardboard sign—frequently rained on and defaced in subsequent days—that said PRESS and hung it on a table in Zuccotti Park. On day 11, September 27, came a breakthrough—a *Newark Star-Ledger* editorial saying unequivocally, "The nation should listen to this small Wall Street encampment."

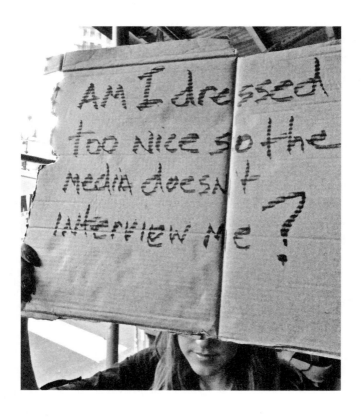

Occupy was not hanging on the mainstream imprimatur. The movement had its own media. A media
team called Global Revolution provided a live stream
video feed. One key member, Vlad Teichberg, a thirty-
nine-year-old Russian immigrant who had walked out of
a lucrative job creating and trading derivatives on, yes,
Wall Street, moved to what he described as "a hard-core

anarchist punk house" in Brooklyn, thrown himself into antiwar video operations, gone to Madrid in the spring to do media with the *Indignados* encampment, and was rigging laptops so they could function as the movement's own video cameras. "The live stream is, in a way, the central nervous system of the entire operation," organizer Max Berger told *New York's* John Heilemann. "Because in moments where the police have tried to fuck with us, it's our first line of defense. And it's been a big part of how we disseminate our information, raise the money, everything."

And, suddenly, there was another energy center in America. Thousands of demonstrators had aroused themselves from their somnolence or despair or plain inertia and poured into hundreds of occupations in public places in big cities, and then smaller ones. They were without any evident organization or leadership, without any headquarters, stationery, or press conferences, and they erected tents, collected food, served thousands of meals every day (more than three thousand in Zuccotti Park alone), organized medical care, transmitted video, and conducted their business in strange human-microphone circle formations (having been denied the use of electric amplifiers, one cop even having barged into Zuccotti Park on the fourth day of the occupation to knock a man to the ground and arrest him for the crime of using a megaphone), wherein anyone could speak, each spoken phrase or short sentence being repeated by as much of the crowd as could hear it. Indefatigable, spunky, self-organizing, and, occasionally, self-disruptive, they formed groups to take up functional tasks. They put together kitchens, libraries, mental health squads, and media teams, published a newspaper, and spun off local civil disobedience actions. They started coordinating with other sites through free conference calls as well as the rest of the communication systems that have become standard in protest worldwide: Facebook pages, Twitter tweets, YouTube videos. Their ecumenical spirit radiated from this call:

> *Occupy Your Life. Occupy your occupation! Whether you clean houses, sit behind a desk, teach in a classroom, work in a kitchen, play an instrument, speak a second language, are a whiz with budgets, can pull projects together and make things happen, make videos, walk dogs, anything . . . your skills are needed at your local occupation square!*

Some of the Occupy pilgrims, as we shall see, carried some fairly sharp ideas about the kind of society they would like to live in (participatory), and reforms that would help (restore Glass-Steagall, tax Wall Street transactions, end the Fed, and public funding of elections among them), and an unwavering sense of the means for achieving either reform or revolution (a mass nonviolent movement). Some brought attitudes painted on signs: "I Hate Drum Circles But I Hate Corporate Greed More," "I Choose to Be Aware," "You Can't Own Land, It's Just as Silly as Owning the Stars," "99 to One: Those Are

Great Odds." Many arrived with little more than a buoyant fury, a prickly sense that the time was ripe (for something), and a summons to lay hold of the world with their own two hands. Their political views were frequently in rapid evolution, as if captured by time-lapse photography. Some, having campaigned for Barack Obama or John Edwards in 2008, might have been occupying Wall Street in 2009 had they not held back, giving Obama the benefit of the doubt, or dazed by disappointment. Their patience had finally worn out, so that they were no longer feeling so thrilled that No-Drama Obama was, when all was said and done, superior to the Republicans. Now, said Jeremy Varon, a historian of the left who teaches at the New School, as he gazed around Zuccotti Park early in October: "This is the Obama generation declaring their independence from his administration. We thought his voice was ours. Now we know we have to speak for ourselves." In the interim, some recent Obama supporters became convinced that the political system was wholly closed, wholly impermeable, wholly controlled by the 1 percent.

It would be a mistake to assume that the radicalism of these improbable world-changers automatically set them against electoral politics once and for all. Hefty thirty-two-year-old high-school dropout, autodidact, and *Daily Kos* blogger, Jesse LaGreca, who declared himself affiliated with the Democratic Party, just before heading down to Zuccotti Park posted these incendiary words: "If I light myself on fire, do you think these bastards will notice?" paying tribute to Mohamed Bouazizi, the Tunisian fruit vendor who did just that and ignited the Arab

Spring. Two months and many public interviews later, LaGreca was still declaring: "At the end of the day, everybody should occupy a voting booth and be a voter."

On peak activity days, the park called in reinforcements. This did not happen automatically: Occupy organizers such as Max Berger and Yotam Marom approached unions, MoveOn, and other membership organizations that could turn out big numbers. Think of the occupiers as an inner movement, numbering somewhere between ten and fifty thousand nationwide, joined for special occasions by an outer movement consisting of union members and progressive supporters (numbering in New York City in the tens of thousands, at times) who gathered for marches that were nationally and internationally coordinated by no publicly known organization and led by no known leaders, and were also willing to show up to help or protect the camps (for example, ringing Zuccotti Park at the crack of dawn on October 14, 2011, summoned by text messages, when the city had threatened to clear the park for cleaning). There were a total of perhaps three million Americans who self-identified with the movement by "liking" one of the 680 Occupy-related Facebook groups and pages. (These numerical estimates come from the Occupy Wall Street stalwart Shen Tong, who was a student leader in Tiananmen Square in 1989, made his escape to the United States, became a software entrepreneur, and then put aside his business to work full-time in the Occupy movement.)

Who led them? "Take me to your leader," big city mayors implored, even as they sent officials to talk with the encampments' negotiating committees, how-ever nonfamous their members. "What do these people want?" journalists wanted to know, even as the loudest, most sustained chant during the marches was unmistak-ably, "We . . . are . . . the ninety-nine percent!" followed by, "Banks got bailed out, we got sold out!" The encamp-ments were consistently unwilling to make the effort to coalesce around what would conventionally be called demands and programs. Instead, what they seemed to relish most was themselves: their community and esprit, their direct democracy, the joy of becoming transformed into a movement, a presence, a phenomenon that was known to strangers, and discovering with delight just how much energy they had liberated. For, indeed, in a matter of days their sparks had ignited a fire.

The collision between the 1 percent and the 99 percent—or those who declared, at least, that they stood for the 99 percent—became a staple of folklore and popular culture. Sometimes the two camps glared at each other. Sometimes they mixed and chatted more or less amicably in Zuccotti Park—Vlad Teichberg was not the only former Wall Streeter who went so far as to switch sides—but for their part, too, most financiers were wary, dismissive, and resentful. A *New York Times* reporter wrote that Wall Streeters experienced Occupy as "morale-crushing." The movement had "made a villain of a once-lionized industry," so why shouldn't wealthy investors return the favor? "A ragtag group looking for sex, drugs, and rock 'n' roll," one top hedge fund manager called them. "It's not a middle-class uprising," sniffed another veteran bank executive, "it's fringe groups." " 'Who do you think pays the taxes?' said one longtime money manager." Had he been Anatole France, he might have added that the law "in its majesty equality" taxes people who take home fortunes more—even proportionately more—than people who have no homes at all, just as it forbids them all equally "to sleep under bridges, to beg in the streets, and to steal bread."

2. Surprise! Popularity and Polarization

The encampments swelled and multiplied. Foreign media paid attention. One week in, there appeared the *New York Times'* longest report to date, Genia Bellafante's "Gunning for Wall Street, With Faulty Aim," which began: "By late morning on Wednesday, Occupy Wall Street, a noble but fractured and airy movement of rightly frustrated young people, had a default ambassador in a half-naked woman who called herself Zuni Tikka." Despite such condescension, or perhaps even recoiling against it, Occupy gathered momentum. More local outlets sent reporters and photographers. Within a few weeks, CNN was running on an on-screen box: "Occupy News."

They launched marches, sometimes down crowded streets, using handheld electronic maps to tell them where to create maximum disruption. Groups dispatched themselves to close their accounts at too-big-to-fail banks. They occupied branches. They crowded into housing courts to support former homeowners who were getting foreclosed. They sent pickets to the mansions of CEOs. They squatted in homes being foreclosed. Collaborating with more conventionally organized groups, especially unions and MoveOn.org, they organized marches of tens of thousands. Each time their secret weapons were the most violent police. Each pepper-spraying, tear-gassing, and billy-clubbing by men in uniform whipped around the world via cell-phone cameras and became a recruiting poster. NYPD Deputy Inspector Anthony Bologna's penchant for pepper-spraying at least two peacefully demonstrating women point-blank in the eyes was recorded in full viral video, on September 24, a gift to what Will Bunch called "journalists who didn't have a clue on what to write about the revolutionary dreamers and schemers who arrived on September 17 and their non-specific grievances" but "did know how to cover an arrest report and to make news out of a video showing evidence of police brutality." Liberty Square doubled its population overnight. Occupiers were astonished by the numbers flocking in.

The police kept coming to Occupy's rescue. But a big reason for the movement's rapid and continuing takeoff went widely unnoticed in the hue-and-cry about how Occupy did or didn't resemble past movements. Unlike any other movement on the American left in at least three-quarters of a century, this movement began with a majority base of support. It was surrounded by sympathy. What it stood for—economic justice and curbs on the wealthy—was popular. People might disagree with Occupy's tactics. They might dislike the occupiers themselves—might disparage them for vagueness, programlessness, leaderlessness, unseriousness, hippiness, disruptiveness, even foulness and smelliness. They might cringe at the riffraff and associate the movement with outbursts of violence even when almost all the violence came from the police. But, even when the movement was seen darkly, through a media lens that highlighted the melodramatic, the garish, and conflictual, outsiders had a pretty clear idea of what was at stake. They could grasp by way of a first approximation what the contrast between the 1 percent and the 99 percent referred to. They could see that the point of the movement was to resist the grotesque inequalities that have become normal in American life, and along the way, to point accusing fingers at a small segment that bore significant responsibility for the economic catastrophe, those who had rigged the game and escaped with impunity.

Baseline popularity would cushion the movement against slings, arrows, billy clubs, pepper spray, evictions, and the occupations' own flaws and failings. Popularity was not everything for a movement—it might just suggest a movement dissolving in a warm bath—but it was a very large something. And it was unusual. How unusual becomes crystal clear via a comparison to the decades usually enshrined as the left's glory years—the sixties and the thirties.

A fog of wishful retrospect has rewritten the recent history of American movements to make them look more popular than, in their time, they were. It accords with the now well-nigh universal lionizing of Martin Luther King, Jr., for example, to think that the civil rights movement was far more popular than, in its time, it was. Obviously it did not go over well with white Southerners in general, but it was not just in the Deep South that the movement was suspect. In September 1960, for example, a national Roper poll asked: "In which direction do you believe the next administration should go on the question of civil rights for Negroes?" Fifty-eight percent of the public chose the middle option: "Proceed slowly until we have worked out the problems resulting from laws already passed." Only 14 percent answered: "Keep on pressing for further civil rights legislation." Likewise, early in August 1963, Gallup asked a national sample of Americans if they had heard of the impending August 28 Washington rally for jobs and freedom. Seventy-one percent said they had heard of it. Those, in turn, were asked how they viewed it. Only 23 percent said favorably. Sixty percent felt unfavorably, either for general reasons, or because they thought it would lead to violence, or because they thought it wouldn't accomplish anything. This is the rally at which Martin Luther King, Jr., delivered his "I have a dream" speech, which is now (at least in the sound-bite form to which it has been reduced for annual consumption) canonical.

As for the antiwar movement, in 1965, when Lyndon Johnson expanded military action in the air and on the ground in Vietnam, his decisions were popular. In the course of that year, support for withdrawal fluctuated between 7 and 20 percent. In seven of eight surveys, the percentage who wanted to expand the war exceeded the percentage who wanted to end it. This was the climate in which the first national antiwar demonstrations took place and the movement grew.

The women's and gay movements also, at their inceptions, met with considerable hostility from both journalists and the general public. For example, according to a Harris survey in 1969, 46 percent of those surveyed deemed working women with young children a force "more harmful" than "helpful to American life." In 1970, only 28 percent of men thought women were being discriminated against in pay, as against 65 percent who thought the opposite. As for attitudes toward gays, in October 1971—more than two years after the Stonewall uprising—fewer Americans (39 percent) thought "homosexual relations with a consenting adult" acceptable than "conform[ing] to society's standards" (48 percent). Both the women's and gay movements had to contend with attitudes of this sort even as they launched their direct actions, trying to change opinions, policies, and deep cultural mindsets all at once. But surely hard times are different. Surely an economic collapse creates circumstances so special as to rope majorities into a collective and conceivably progressive consciousness.

So, then, is it *economic* grievances—the ones that bite into majorities—that automatically bring public support for left-wing policies? Not so fast. There would seem nothing automatic or predictable about the flows of American opinion. It's interesting to note, for example, the ambivalence with which the public met the industrial union drive of the 1930s. On the one hand, during the second half of the decade (there was no good polling during the first half), three-fifths of the public said they either liked Franklin Roosevelt's attitude toward union labor or thought it insufficiently friendly. A full 79 percent agreed that workers had the right to join unions. But even then, 75 percent thought unions should be licensed by the government, and 70 percent thought employers had the right to ask workers whether they belonged to a union—a policy that does unions no good. At a time of intense conflict between the Ford Motor Company and the Automobile Workers Union over the workers' right to organize, between twice and three times as many Americans (depending on how the question was asked) sided with Ford as with the union. Almost twice as many Americans preferred conservative AFL-style craft unions to militant CIO-style industrial unions. Even as the right to form a union was being fought for and died for, even as one-quarter of the work force was unemployed, America wanted the right to unionize, sure, but was also suspicious of unions.

So it is all the more striking that OWS, from the start,

should have floated so high on a wave of popular sentiment. Even before the Zuccotti Park occupation, polls
consistently showed supermajority support—60 percent
and more—for progressive economic reforms like raising taxes on households that earn $250,000 per year.
Seventy-six percent of the public favored increasing the
taxes of people who make more than a million dollars
a year. Asked what their greatest economic worry was,
nearly twice as many Americans named jobs as named
budget deficits. According to one November poll, six in
ten Americans said they supported government efforts to
reduce disparities in wealth. A different November poll
showed 60 percent of registered voters "strongly agreeing" with this statement:

> The current economic structure of the country is out
> of balance and favors a very small proportion of the
> rich over the rest of the country. America needs to
> reduce the power of major banks and corporations
> and demand greater accountability and transparency.
> The government should not provide financial aid to
> corporations and should not provide tax breaks to
> the rich—

as against 33 percent who preferred to cut the national
debt, reduce government, deregulate, and oppose new
taxes. Plainly, the movement was perched on a cushion
of popular support.

Or was it? Even optimists should proceed with care,
not get carried away by numbers, even the cheery ones.

Poll results can be slippery under the best circumstances. People may respond to pollsters as they think they're supposed to respond, feeling pressed to have opinions about questions they really care little or nothing about. And they contradict themselves, if for no other reason than that the questions are so often simple-minded. That said, it still gives one pause that the same poll showing six in ten supporting government efforts to reduce disparities in wealth also showed 56 percent favoring a "smaller government that provides fewer services" over a larger government that does more. There were as many respondents (33 percent) who strongly agreed "that the defining problem with the economy today is too much government" as those who strongly disagreed, while another 20 percent "somewhat agreed." If people wanted more equality but less government—presumably meaning less taxation and less regulation—what did they really want? The support for Occupy's goals that the polls picked up might well have been softer than, at first blush, it appeared. And if Occupy was incoherent, what would we call the American public?

Public self-sabotage is embedded deeply in American culture. Not for the first time, a goodly number of Americans wish to have their freedom and eat it at the same time. They want to will an end (more equality) while refusing to will the means (reform via government action). Partisans of equality—at least the idea of greater equality—are loath to risk surrendering freedom, which to them means freedom from government. Whether they know it or not, they are stuck in the jaws of ambivalence. But the American people are the American people. They cannot be dissolved and a new one summoned.

Accordingly, it should not have been surprising that, in its second month of existence, the Occupy movement was butting up against headwinds, even as its goals, at least their general spirit, won favor. As the movement became better known, most people made its acquaintance through the lens of the media, which were, as always, drawn to the most spectacular images of clash and confrontation. Even the movement's own media shared the same fascination with the drama of human beings colliding with other human beings and inflicting harm on them. Not surprisingly, the public increasingly associated the movement with violence and the disruption of everyday life.

Reports of clubbings, arrests, and evictions were newsworthy—easily translated into action stories, headlines, arresting photos. To journalists, singling out such events was a no-brainer: The stories wrote themselves. Never mind that the demonstrators were, all told, less violent than football fans. When the police shoved demonstrators, clubbed them, and gassed them, the scenes of the action were dubbed "violent clashes," as if nonviolent demonstrators were responsible for police attacks. But this was the normal way of media, which know a good deal about how to attract attention from persons who are not normally attentive. In November and December, the paramilitary police forces of at least eighteen cities used pepper spray, batons, and even sonic grenades to evict Occupy activists. Protesters plausibly suspected coordination. And, indeed, the mayors of Oakland and Portland, Oregon, publicly confirmed that eighteen mayors had taken part in a conference call beforehand, and a Justice Department official claimed "each of the Occupy raid actions were coordinated with help from Homeland Security, the FBI, and other federal police agencies." As the grander encampments were cleared out, militant actions snowballed. Activists, after all, strived for momentum. Grow or die, show up or die—these were de facto mottos of all movements. "As we have learned," trenchant Berkeley blogger and Occupy Oakland activist Aaron Bady wrote in defense of the militant turn, "peacefully setting up camps—in one of the least confrontational forms of civil disobedience imaginable—is something that American cities will simply not allow. So what is left to the Occupy movement but more aggressive and confrontational tactics? Since they are not allowed to camp peacefully in public parks, expect more work stoppages, more foreclosure defenses, and more building occupations."

Accordingly, the country polarized. The more people heard about what Occupy said and did, the more they divided. A little detour on polls will make the point and also show how fuzzy it is. One poll showed that between mid-October and the time when Zuccotti Park was cleared out, approval of Occupy *tactics* sank from 25 to

20 percent, while disapproval rose from 20 to 31 percent. In other polls, those who called themselves supporters of Occupy sank from 29 to 24 percent, and those who rejected that label rose from 62 to 68 percent, between early November and early December. On the third hand, also in early December, Pew found far more Americans agreeing with "the concerns raised by the movement" (an awkward formulation, but still) than disagreeing (48 percent to 30 percent), while nearly identical numbers disapproved of Occupy *tactics*.

What were such numbers good for? The closer one looked, the more blurred the tea leaves. It might be that opinion about Occupy was soft, greatly dependent on precise wordings and unaccountable fluctuations of public moods. In the event, what the polls said about Occupy was no more unreliable than what it said about other phenomena, so it is interesting that during its third month, Occupy pulled even with the Tea Party, which was then in its third year. The two movements were more or less equally popular—and equally polarizing. They mustered similar proportions of strong supporters, with a slight edge to Occupy (18 percent, to 14 percent strongly in favor of the Tea Party). What to make of this? In important ways, the two movements are not symmetrical. Popularity may matter more to the Tea Party, whose thrust is largely electoral, than to Occupy, which is at one and the same time more disruptive and more utopian. Popularity is not the be-all and end-all of a growing movement. There are times when it may make sense to opt for divisive tactics—to keep up momentum, to recruit more ardent followers, to convince elites to make concessions—even at the price of alienating large swathes of the population. Defenders of confrontational tactics may legitimately point out that popular movements do not grow smoothly, either in core numbers or in the larger circles of support they activate. Still, a movement's cavalier indifference to (or even taste for) growing popular antagonism suggests a toxic combination of recklessness and insularity.

Cautionary case in point: the fateful paradox of the turn toward confrontation politics in the late sixties, when even as the Vietnam war grew steadily more unpopular, so did the antiwar movement. Most Americans abhorred disorder to begin with. They liked comfort and convenience. When militants filled streets that commuters were trying to drive through, the commuters were less than enthusiastic about "the revolution." Making their arduous ways through everyday life, ordinary people preferred not to be interrupted, delayed, reminded of the shaky ways of the world. There was this factor, too: As many grew to detest the war, they also hated to be reminded how awful it was—most of all, of course, when the agitators doing the reminding were such irritating, arrogant, moralistic hotheads.

For their own part, fervent and euphoric antiwar militants were in the grip of a go-it-alone attitude for which they found ideological justifications. Intoxicated by their militancy, thrilled to confront the authorities (*any* authorities, with the police often enough ready and eager to fill the parts for which they'd been virtually recruited by Central Casting), taking pleasure in tactical innovations, they were fully capable of shrugging off popular opposition as proof of their rectitude. (At the outer limit of their immense arrogance, the Weathermen splinter of Students for a Democratic Society liked the ring of the slogan, "Fight the people.") They could rightly point out that popularity was not power. If they had chosen to make the case, the militants could legitimately have claimed credit for a big success—pushing governing elites toward policy change by the sheer force of their disruptive power. At decisive moments in the late sixties and early seventies, they massed large numbers to forestall and reverse huge escalations of the war. But the desperation and foolhardiness of go-for-broke activists, who had devoted fervent years to the antiwar cause and still failed to stop the slaughter in Southeast Asia, also held enormous and unhappy implications for the movement's future. Richard Nixon was not the only politician to ride the menace of the disruptive left into power, until his paranoid hatred of the antiwar movement led him to overreach, and eventually brought him down.

Forty years later, in the third month of Occupy, life in the encampments grew more agitated and dangerous. Homeless people, untethered by political discipline, were conspicuous. Violence against women was reported. Big-city mayors sent in the police to displace the encampments, generating more militant actions in counterreaction. Occupy's tactics threatened to overshadow the movement's ideological thrust, its import, its political meaning. Some supporters worried that America was going to repeat itself. Would it approve of the Occupy message and come to despise the messenger?

3. Energy Is Eternal Delight

Deflated, aghast, grumpy, and largely inert, the American left had watched helplessly as the Obama wave of 2008 subsided, its grass-roots organizing apparatus disbanded, and young voters who imagined that the professor-politician from Hawaii and Chicago was in his own person, despite all his denials, the embodiment of change, and that they had discharged their political duties by knocking on doors to elect him, returned to their usual pursuits. In the meantime, from the beginning of the new administration, amplified in their media echo chamber, the tax-obsessed Tea Party had burst upon the national scene and gone on to dominate national politics for the better part of two years, even as unemployment was locked into a disgracefully high rate (around

9 percent in official figures, but well above 15 percent if one counted those who were no longer looking for work and those working part-time because they could not find full-time work). The result was that vast numbers of the young voters of 2008 had stayed home from the midterm elections of 2010, with a predictably dire outcome for the Democrats. In 2008, voters under thirty had turned out at the same rate as voters over sixty-five—if you were looking for a single-factor demographic explanation for Obama's victory, you need not have looked further. In 2010, the youth turnout had melted back to its norm—somewhere between one-third and one-half of the seniors' rate—and predictably the Democrats had been crushed. Meanwhile, the perpetrators of the financial collapse, and its chief beneficiaries, had walked away virtually unscathed.

Now, as the encampments endured and grew, a demoralized left shook itself, stood up, and found to its amazement not only that it existed but that it radiated energy, and that energy, in the social world as in the material world, made things happen. Its dynamism nudged some who were troubled about the consequences of the economic crisis, but noncommittal about knowing what to do about it via progressive remedies, despite the movement's own diffidence about demands. It persuaded some whose thinking had migrated leftward to act on their own. It moved some occasional activists toward a more intense commitment. It energized those who were already committed. And it produced its own satisfactions, since as William Blake understood, "Energy is Eternal Delight." (See part II, The Spirit of Occupy.)

Within a couple of months, sympathetic pundits and politicians were agreeing with journalists that Occupy had changed the debate in the country, had put inequality on the agenda. "The whole battleground has changed," Senator Charles E. Schumer of New York, the top political strategist among Democrats in the chamber and also their star fundraiser on Wall Street, said in an interview. "There's been a major shift in public opinion. Jobs and income inequality are going to be the number-one issue" in 2012. Not surprisingly for a senior senator from the state of New York, given the immense costs of running for influential office, Schumer was himself a master of the fine art of maintaining a liberal reputation while raking in Wall Street dollars. In the previous election cycle, Schumer's intake from the sector known as FIRE—Finance, Insurance, and Real Estate—amounted to $5.6 million, totaling 29 percent of his total intake (nearly $20 million) for the period, and by far the largest sector from which he drew his support. The atmosphere was such that, in Albany, Governor Andrew Cuomo changed his position on the so-called millionaire's tax. Previously, he had proposed that this surcharge on single individuals making more than $200,000 and married couples with incomes over $300,000 lapse. Suddenly, in December, he signed into law a bill that would reinstate a higher bracket for households with incomes over $2 million per year, although (as many journalists failed to notice) that bracket was still lower than it had been before the surcharge was imposed, and many millionaires would get bigger cuts than families earning less than $300,000. Meanwhile, several banks that had raised fees for debit cards and other products rescinded them. The climate had turned—at least a bit—against the impunity of the wealthy.

Even before Occupy surfaced, the Tea Party was receding in popular support. According to one poll, the percentage of respondents who viewed it unfavorably rose by half, from 26 percent just before the 2010 midterm elections to 40 percent in August 2011. Having zoomed into the media spotlight during 2009 as the only populist movement in sight, and spread whip-fast, combining an antitax backlash with anti-Obama rage during the months when Obama was squandering the momentum with which he had arrived in the White House—a Tea Party largely made up of older, whiter, more prosperous, more educated, more male, more married, and largely evangelical Christian Americans benefited from media splashes, accrued the largesse of billionaires, and won the fervent support of a Washington-based right-wing apparatus controlled by the former Republican House Leader Dick Armey. While the left's street activists, busy promoting health care reforms, were consigned to a scatter of local news reports, the Tea Party ballooned into *the* national story about political opposition. According to statistics blogger Nate Silver, Tea Party coverage mushroomed by a factor of six during the three weeks that followed their tax protest media launch on April 15, 2009.

In January 2010, Scott Brown, a handsome state senator who drove a pickup truck to maximum advantage, won the Massachusetts special election to fill Ted Kennedy's seat, and his victory was heralded as a Tea Party breakthrough. Then, rightly or wrongly, the Tea Party was credited with responsibility for the Republican gain of sixty-three Congressional seats in the 2010 midterms. But, then, a good many Americans who had rallied to their antigovernment rhetoric found a Tea-infused Republican Party less fragrant once it drove the country to the brink of debt default during the summer of 2011. Once in office, Tea Party representatives refused to raise taxes on millionaires and showed no compunctions about tossing government employees into the gutter. Barack Obama was taking a lot of flak from his left for steadfastly performing the role of reasonable man trying to broker compromises between Democrats and Republicans, but his strategy finally paid off: The Tea Party movement was cast as wreckers and obstructionists.

As the Tea Party's glamor faded, Occupy replaced it as the big political trend story. On October 1, more than seven hundred marchers were arrested on the Brooklyn Bridge, a photogenic event partly contrived by handcuff-happy cops who lured some of them onto the roadbed, and partly contrived by marchers who wanted to be precisely there, euphorically blocking Saturday night traffic, in an event readymade to lead the evening news.

Four days later, with the media paying much more attention, that civil disobedience moment and dutiful police overkill was followed, in turn, by a lower Manhattan turnout of ten thousand or more not noticeably scruffy demonstrators on October 5, along with demonstrations in more than a hundred other cities across the country, which now, three weeks in, recast the movement first as innocent victims, then as ordinary citizens, and overall as a widespread and growing bloc of apparently sensible, normal-looking people who raised reasonable alarms, made sensible proposals, and enjoyed popular support. ABC's *World News* treated the national demonstrations as a major story that night, featuring one of those animated maps on which cities light up all over the country. The foreign press, interested from the start, showed up in throngs. Reporters from three different Polish newspapers showed up on October 5, along with the BBC World Service, as well as the American networks. A Dutch television outlet followed a Transit Workers representative around for the day—for the TWU had protested their buses being commandeered to drive Brooklyn Bridge arrestees to jail, a story that interested the Dutch more than it did American mainstream media.

Accordingly, the media preoccupation with the righteousness of cutting government budgets faded, and the austerity caucus lost its lock on the public agenda. In the course of the fall of 2011, according to one Nexis survey, the media focus lurched over to income inequality and disparities of wealth. Within seven weeks, the term *income equality* was cropping up in the Nexis database almost six times as often as it had during the week before the encampment settled Zuccotti Park. For months, Washington and the media had been riveted by the dangers of government spending, far more than public opinion was. Now the mainstream media—the country's semiofficial playlist of legitimate themes—caught up with the kitchen table. The unspeakable was speakable.

Whatever imprint the movement leaves (or fails to leave) on national life, this spectacular uprising, within a bare few months, accomplished one of the prime objectives of any social movement: It upturned millions of people's sense of the possible. A hard core of a few thousand activists, who occupied (camped, agitated, and so on) more or less full-time, had stirred tens and hundreds of thousands of others to play other essential parts: to donate money, granola bars, muffins, and apples, and order pizzas for delivery from local joints (including, a couple of blocks away, Liberatos—its actual name—which after granting Occupy permission to link from its website, invented the $15 OccuPie, with a diagonal pepperoni stripe signaling "no," and soon was sending sixty pies per mealtime, including vegan specials); to help with legal and medical and public relations expertise; to shut down their big-bank accounts and make political demands either explicitly or implicitly; to join the big marches, coordinated among a growing number of cities via free conference calls and social networks; and, in a thousand ways, find, or create, a piece of the movement for themselves. In all, what began was an awakening.

There were new players in the game of changing the world, occupying a new center of social initiative, and millions felt its pull. It seized attention, won friends, enlivened the demoralized, stirred up theater, engaged wits, penetrated the media as a big cultural and political fact, influenced some people, antagonized others (not least America's increasingly militarized police forces and mayors who feared political backlash if they let the encampments go on), and prompted intense conversation not only about the financial crisis and about causes of and remedies for the deep economic affliction, and about larger, chronic injustices, but about what ought to be done not only by political leaders but by everyone within earshot.

How was such an uproar possible so fast? How did "Occupy," "1 percent," and "99 percent" become everyday terms, Zuccotti a tourist site, and pepper spray an instantly recognizable emblem of thuggish policing? And why was SnoringCenter.com buying billboard space in Chicago to advertise "OCCUPY YOUR BEDROOM: DON'T LET SNORING KEEP YOU APART"?

It's a commonplace that we now live in an era when media—global and local, official and unofficial, home-screened and hand-held—speed up the cascades of events. The movement's own videos, slogans, and slogans went electronically viral from New York to California, revealing a lurid profusion of unnecessary force, not least the profligate use of pepper spray. Even mainstream media disclosed violence perpetrated by police against demonstrators, from the firing of rubber bullets to the fracturing of the skull of ex-Marine Scott Olsen as he took part in a peaceful protest at Occupy Oakland on October 25, 2011, to the astounding casual pepper spraying of conspicuously nonviolent, nonthreatening students sitting down on the campus of the University of California, Davis, protesting tuition increases. For it was no small part of this movement's achievement that Occupy's thousands hardly ever resorted to violence themselves, and, therefore, continued to win the battle of theatrics. Video provided stark evidence of Lieutenant John Pike spraying unresisting Davis students sitting calmly on the ground, having refused to take down their tents—to which, as the video demonstrates, the shocked crowd reacted with nothing more (or less) than the chant, "Shame on you! Shame on you!" followed by calls of dismissal, "You can go!" The students were Gandhian without need of a Gandhi. The exceptions, at this writing, were occasional and unsystematic, at the movement's edge, the most prominent being a bout of window-smashing in downtown Oakland eight days after the wounding of Scott Olsen, when about a hundred people in black masks and bandannas broke from a largely nonviolent demonstration while many among the large majority tried to stop them.

Angry and exuberant, inventive and chaotic, confusing and polymorphous, this movement came roaring out of nowhere into public view and then outward, via vast webs of communication, beyond just the movement's own, into the nation's living rooms and around dinner tables, making it legitimate—more legitimate, anyway—to ask the left's classic question: What is to be done about flagrant inequalities? And another: How did it come to pass that 1 percent of the American population controlled 40 percent of the wealth? And: What happens to the American dream when the middle class (which, whatever that may be, is the category that most Americans identify with) is squeezed, and many millions doubt they will be able to live half as comfortably as their parents? And: Is there a moral imperative in American life besides *Enrich yourself?*

These were especially urgent questions, of course, for millions of people without work, or buried under debt, or up against the prospect of foreclosure and eviction, or the younger ones who had packed up and returned home to live with their parents and now found themselves staring into a future of improvisational freelance or otherwise interrupted work. It was not that the millions of people who sympathized with the Occupy movement had suddenly discovered that they were aggrieved and fed up. On September 16, 2011, the unemployed already had sick feelings about being unemployed, or seeing close family members unemployed, and many of them could give an account of how this had happened to so many, though they might disagree about how much blame to allocate among the investment banks and mortgage brokers, collusive government agencies, super-profitable cash-hoarding corporations and their self-dealing CEOs, paid-for elected officials, and the rest of a far-flung, intertangled system of Wall Street and Washington back-scratching that was properly called *crony capitalism.* They were disgruntled and indignant and anxious in the short, middle, and long runs, whether on behalf of themselves or their families and friends or, more encompassingly, on behalf of an entire nation gone sour with curdled dreams. What was eerily absent—given the disappointments after Barack Obama took office, and the Tea Party's seizure of the political initiative—was the energy to reach down into their sense of outrage, to distill from it a sense that something better was possible, and to devise forms of do-it-yourself action that would enliven a critical mass.

PART TWO
THE SPIRIT OF OCCUPY

4. Oases and Bases

If you visited an Occupy encampment, you might think you had stumbled across a lost tribe—or a found tribe, possibly one made up of time travelers, plunged unaware into the present, leaving it to you to judge whether they are utopian or dystopian, or simply odd.

A circle sits, or stands. Somebody calls out: "Mic check!" Without hesitation, as if they were only awaiting the signal, some people within hearing range repeat: "MIC CHECK!" The sequence repeats, the echo louder this time: "Mic check!" "MIC CHECK!" "We have a problem here—" the speaker might say. "WE HAVE A PROBLEM HERE—" "Look around." "LOOK AROUND." "We're a circle of palefaces." "WE'RE A CIRCLE OF PALE-FACES." "We need outreach into the neighborhoods!"

"WE NEED OUTREACH INTO THE NEIGHBOR-HOODS!" and so on. Periodically, a scattering of people more or less simultaneously raise their fingers into the air and flutter them. "Twinkling," this is sometimes called. Others lower their fingers and wiggle them:"de-twinkling." If you watch long enough, you get the point: Twinkling is the equivalent of applause. De-twinkling is the equivalent of a hiss. People raise their hands to get their names placed "on stack"—on the speakers' list. Occasionally, someone raises both hands above his head and forms a triangle with his fingers. The triangler might be noticed by another person, who stops the discussion and calls on him. The person with the authority to stop the conversation is a facilitator. The triangler is raising a procedural point. More uncommonly, when a proposal is put up for adoption, someone crosses raised fists and forearms. This is a block. It stops things cold.

Meetings go on for hours this way. The subgroups that carry on most of Occupy's activities, Working Groups, operate in the same manner, even in meetings of over a hundred. There are ninety-seven of them thus far—Think Tank (the largest, with 794 members signed up online), Direct Action (719), Arts & Culture (664), Politics & Electoral Reform (572), Tech Ops (552), Vision & Goals (519), Facilitation (439), Outreach (423), Empowerment and Education (374), Movement Building (331), Alternative Banking (303), Structure (283), Political Action & Impact (273), Finance, aka Accounting (250), WOW, Women of Wall Street (246), People of Color (229), Environmentalist Solidarity (222), and

Legal (211), to mention only those numbering two hundred members or more. As one can imagine, there are jurisdictional disputes, sometimes bitter ones. But, at least sometimes, the meetings pause for jokes, sometimes on themselves, and people laugh uproariously. ("How many anarchists does it take to change a light bulb? Don't change it, smash it." "How do you know there's a hippie in your house? He won't leave.") These are the notorious legions of Occupy who bring mayors to their knees when they talk about pitching tents on a piece of ground, or in a bank, say, or march to the Brooklyn Bridge, and sit there in a circle, repeating their phrases, making their funny gestures.

One of Occupy's press attachés, thirty-two-year-old dark-haired, angelically smiling Shane Patrick, who grew up in Queens, the son of Irish immigrants, is a veteran of years of left-wing activities in and around lower Manhattan. The experience wasn't always wonderful. He knew, for example, those times when a sectarian group turns on you, having decided that you are deficient in revolutionary zeal, and then a comrade whom you shared a jail cell with after a militant demonstration a few months before wants nothing more to do with you anymore, and when he runs into you on the street, turns away. Patrick had been through that wringer. So, when he started coming around Zuccotti Park, around Halloween, he felt the absurdity of the GA rigmarole. "It feels like a *Monty Python* sketch," he says. "And you're feeling you're the only one in the circle who feels that way, who gets it. But you're not."

Mark Bray, a twenty-nine-year-old history graduate student who has been a political organizer since he was seventeen (antiglobalization, immigrant rights), found out about the September 17 rally on Facebook, marched to Zuccotti Park on day one, but wasn't so thrilled with the superficial breakout discussion he ended up in, which concerned what a single demand—if there was to be one—would look like. End corporate person-hood? Make tax rates more progressive? Such slogans, he thought, were not inspirational. He picked up some Chinese food and headed home.

His trajectory then tracks the early dynamic of Oc-cupy and illustrates how it attracted so much talent so quickly. Over the next few days, some Facebook friends posted about Zuccotti developments, so Bray kept up. He noted that the occupation continued. A few days in, he saw a video "of an occupier being dragged out of the park by a police officer without any evident justifica-tion, and despite my misgivings about Occupy, I felt for those who were at least doing something." The fact that the encampment lasted made him curious. He went on some marches. "Despite the odds, they were getting out there and not caring whether they had ten people or five hundred." He saw the YouTube video of Officer Anthony Bologna pepper-spraying two young women on the side-walk, which outraged him. On September 29, he was surprised and impressed to see that the Transit Work-ers Union endorsed OWS. "It became clear that Occupy was striking a deeper chord in society."

The next day, Bray continued, he went on "a really great, spirited march from Zuccotti to Police Plaza. I really liked it because it was enthusiastic in a really fresh way, and nonsectarian. Usually at marches you see all the various socialist parties and factions who are handing out their newspapers, etcetera, and this was just people who were angry, marching with unifying slogans that many can get behind. There was a great feeling of building momentum, of solidarity, and defiance. It was just a great vibe." So much so that the next day, October 1, Bray changed his plans and joined the march over the Brooklyn Bridge. "I had the palpable sense that something really unusual was happening there," he told me. He watched from above as "people held an impromptu GA, right there on the roadway. There was a mic check, to get people to stay calm, and most people sat down and tried to figure out what they should do as a group. By that time, the police were moving in, and as the arrests started, people started to stand up again, and the ability of the GA to function crumbled. But that moment of unity and solidarity under these circumstances was very," he paused, "moving." He'd been thinking of getting more deeply involved, but the march pushed him "over the edge toward being an Occupy organizer. The bond connecting this large group of people was stronger and more authentic than I imagined possible after two weeks of organizing." He threw himself into Occupy's Public Relations Working Group.

The encampments were, while they lasted, bases, but

also oases. Urban tribes (or, in more social-scientific lingo, subcultures) collected, clustered, congregated, were generous, shivered, and sometimes squabbled there. They devised and observed rituals. They hung out and befriended each other (in the non-Facebook sense), applied and learned skills there, argued about the shape of a decent society, organized committees, observed and irritated each other, assumed leadership and were acknowledged for it, and recognized the leadership of others, all in an ostensibly leaderless movement.

Some looked at these placard-festooned tent villages and saw "goddamned hippies," (attributed by Rupert Murdoch's tabloid *New York Post* to an "irate" and "livid" but unnamed Ground Zero construction worker said to have been blocked getting home from work over the Brooklyn Bridge), "Occupy Wall Street animals go[ing] wild" (a *Post* front page headline), "destructive parasites," "morons," and "trust fund hippies." Others saw the tents and felt attracted by what they called the "Woodstock vibe" (though Zuccotti Park was far tidier than Woodstock had been, lacking the mud wallows of 1969, and frequently swept clean by the brooms of the denizens). The sites attracted musicians of many genres, cameras of all descriptions, spray painters of varying talent, designers, librarians, tech experts, religious zealots, friends embracing old friends, and occasional wafts of marijuana, though perhaps no more than on many other blocks of the city.

There was sweetness and affection and gaiety along with exhaustion, but the premium style was *earnest*. The occupiers could be terribly earnest, endlessly debating the rules of correct democratic process or listening to Joseph Stiglitz and Jeffrey Sachs present arguments about inequality. For all the playfulness, most conspicuous in their posters, for all the cheer evident in their improvisations, the camps formed a sort of subculture of seriousness in a sea of snark, that combination of sarcasm and cynicism which has been the dominant tone of the young for some two decades. Snideness was not banned in the encampments, or the working groups that spun off it, but snideness was not the premium style.

They were earnest because of what was at stake, which was no laughing matter, even if the new normal, the accustomed condition to which they refused to become accustomed—which they had inherited and which looked to be their destiny—was SNAFU: Situation Normal, All Fucked Up. (Or, in the words of a Zuccotti Park sign that quickly caught on as a description of a world so badly wrecked as to transcend the very capacity of propositional speech itself to state just how and why: SHIT IS FUCKED UP AND BULLSHIT.) They were declaring with all due gravity not only that their personal livelihoods were at risk but that the same forces responsible for the precariousness of their lives were equally responsible (that is, irresponsibly) for the grave, interconnected hazards that had the world by the throat. One October evening in the East Village, a young anarchist, who had launched his occupation career at the encampment of *Indignados* in Barcelona, looked at me across a bar table and shifted effortlessly from an outburst of mirth to a grave soliloquy: "Look, I'm twenty-five years old. I'm never going to have a real job. And the ice caps are melting." Who would dare tell him to take it easy? Earnestness was the new counterculture.

So, the Occupy camps were Woodstock, suffused with generosity, and they were also nonstop bull sessions and more formal seminars on economics, where newcomers might learn (if they hadn't already suspected from experience) such facts as that unemployment for sixteen-to twenty-year-olds was 24 percent, that one-quarter of American workers were making less in 2011 than the minimum wage paid *in 1968*, and that those with more than a high school education were leaving (or never entering) the workforce at the same rate as those with high school degrees, and that for every job opening there were more than four unemployed people, this ratio not varying much across industries.

As journalists quickly learned whenever they made like Martians in the old post-Sputnik jokes ("Take me to your leader!"), the encampments were organized horizontally, without evident hierarchy. At Zuccotti Park, they received the support of the local community board, but they could not control the drummers who defied the General Assembly's drumming rules and infuriated neighbors, or the stray anti-Semites holding "Jews Control Wall Street" signs. They knew that they needed to face outward, toward the vast stretches of America and the world beyond the perimeters—to "build the movement," to "do outreach," to "expand the base"—but they did not agree on which base was most important, and what to do about the fact that African-Americans and Latinos and other racial minorities were underrepresented, and what to expect of different segments of the 99 percent. Still, the spirit of generosity never dissolved. On Christmas Eve, the Occupy Community Relations working group delivered gift boxes of vegan doughnuts to local restaurant workers, security guards, and others who had helped out with food and bathrooms during the months of encampment.

Around the country, hardly any counterdemonstrations cropped up. While the weather held, a good number of Manhattan tourists riding on the upper decks of buses flashed V signs and solidarity fists. Most passersby were curious, and got friendly receptions. But the encampments were frequently preoccupied with their inner operations. This was, in many ways, a community relating to itself—so much so as to give some in the movement, not just its theoreticians, the feeling that the occupations prefigured a good society, that they supplied, at least, whiffs of utopia, even as critics inside the movement thought there was altogether too much navel gazing going on. "The process is the message," wrote Meghann Sheridan on Occupy Boston's Facebook page. "Sometimes," said one member of the OWS working group, Think Tank, in December, "I think the conversation is one of our goals." "Just being able to say what you feel is one of the most empowering things," said another. They planted themselves amid alien buildings in order to *occupy*—the verb coming to signify an intransitive, as in performing an act that did not require an object, like dancing, not only the transitive, as in occupying a particular space. Many occupiers felt like peaceable warriors, arming themselves with mental weapons to redeem the surrounding desert. The homeless, the vagabonds, the deranged and unruly also took refuge there. To assemble for the redress of grievances could feel like undressing for the wildest sort of release, or intoxication, or like the dressing of wounds.

Here, outsiders repelled by a cold society could feel like insiders in a warm one, even if they had to camp out on cold stone to feel it. The camps, sprawling and apparently chaotic, not infrequently irritating, even crazy in the eyes of outsiders, yet somehow self-regulating centers of public hubbub were outposts of solidarity where people began as strangers to one another and, many of them, felt themselves evolve into members of a commons. They became both communities of self-government and incubators of identity, even as real experience broke the mystique and the camps exposed their ugly sides, the criminality and mental disturbance and ego-tripping. Liberty Square, it turned out, harbored enclaves of license, too. By mid-November, the encampments were, in the eyes of one mental health worker whom I shall call Jennifer Lightfoot, "too intense." Predatory men harassed women. There were heavy drugs. In reaction, there were also lots of histrionics, rumors of violence. An all-night community watch was organized. "People were shooting up," Lightfoot told me. "It was like *The Wire*. There were dealers. There was a gun or two. Somebody would have died," were it not for the eviction carried out by hundreds of riot police in the wee hours of November 15. Lightfoot was not the only one to express (privately) relief about the eviction. In fact, I must have heard this from more than a dozen Occupy people, usually unprompted, and although many acknowledged that communicating had become harder after the dispersal, the dominant note was relief.

By creating their little worlds in Liberty Square and its hundreds of turf equivalents elsewhere, the campers were of course making a statement in the spirit of the First Amendment, availing themselves not only of freedom of speech in the abstract but of "the right of the people peaceably to assemble," a less frequently cited phrase in the same sentence. In truth, they constituted, with their bodily presence, a human petition—an outpouring of conviviality and a declaration of loyalty to the 99 percent and a desire to chastise and curb the 1 percent who prospered during and despite and sometimes because the bottom had fallen out of tens of millions of lives as a result of a foreseeable financial crisis for which they were responsible. That much was obvious to the unblinkered. But to the consternation of journalists who wore blinders to the encampments, thinking that they already knew what a political movement looked like (and these so-called incoherent assemblies didn't fit the bill), but also of well-wishing reformers of (or even revolutionaries against) capitalism run amok, they were there not only, or even primarily, "to petition the Government for the redress of grievances." They had both more and less in mind. They felt they were inhaling the spirit of revolution, overturning calcified structures and upturning new soil. And they felt what they chanted, that "the whole world is watching," just as they had been watching North Africa, the Middle East, southern Europe, and earlier, homegrown Madison, Wisconsin, where union members and their supporters numbering in the tens of thousands had risen up and, indeed, occupied the State Capitol to protest the rescinding of collective bargaining rights by the Republicans.

They were inspired, above all, by the so-called Arab Spring, the previous winter's gift to the global idea of popular power overthrowing dictators (if hardly, as it turned out, all systems of oppression). This was not to say that, in lower Manhattan, they necessarily knew much about North Africa, or that all of them quite grasped the difference between the occasional use of orange nets, pepper spray, and heavy-handed paramilitary police evicting them from Zuccotti Park that some called signs of a police state in New York City, on the one hand, and the rampant disappearance, torture, mutilation, and murder practiced in Zine El-Abedine Ben Ali's Tunisia and Hosni Mubarak's Egypt (and its military successor) on the other. They knew an international inspiration when they felt it. A few of the campers had passed through the Spanish encampments in Madrid and Barcelona during the spring and summer, where they had passed time with *los Indignados*, expressing their disgust with an entire political class that had left one of two young people jobless and the whole society impounded by an austerity with no end in sight. Like the British artisan, agitator, and pamphleteer whose *Common Sense* inspired the American revolutionaries more than two centuries earlier, and who later returned to London and wrote *Rights of Man,* only to be sued for seditious libel, and who fled to France, where he was elected to the National Convention (never mind that he didn't speak French), but voted against the execution of King Louis XVIII and was condemned by the ultras to the Bastille, where he finished *The Age of Reason* and barely avoided the guillotine—like Tom Paine, the English radical turned American revolutionary turned French revolutionary, his successors were globalist. They were in Liberty Square, as an American minstrel sang before most of them were born, "to breathe the air around Tom Paine."

5. Rituals of Participation

"*O*ccupation is more than just a tactic," wrote one of the movement's most articulate analysts, Jonathan Matthew Smucker, just five days before the New York Police Department made another in its series of contributions to Occupy's tactical evolution by surrounding Zuccotti Park, sealing it off from the neighborhood (and from reporters and cameras), and evicting the occupants in the middle of the night. "Many participants are consciously prefiguring the kind of society they want to live in." Not many were so naïve as to think that they had already created utopia—a condition which was named originally, after all, to mean "nowhere." But the talk about "prefigurations" and "models" and "intimations" was not idle, either. It expressed a passion, a will to believe, that it was possible on one patch of ground, even provisionally, even approximately, to plant a foot in a future of active engagement and free expression even (or especially) for those not used to commanding attention—*empowerment*, in a word. "The Occupy movement calls us into the streets, toward public engagement, and to the modern day polis that is the general assembly," wrote Brooke Lehman, a longtime practitioner of the arts of direct democracy. "It calls us with a single unspoken but implicit demand: participate!" Others cautioned against "the backs to the people aspect of the GA," the emphasis on group maintenance in the assemblies that both literally and figuratively closed themselves off from passersby to spend hours each day transacting their business.

The spirit of Occupy thrived, in no small part, because the movement spawned mantras, rituals, symbols, imagery galore—a riot of pastiche and contagion. First but not least, there was the terminology, suitable for infinite multiplication: Occupy Wall Street, Occupy the Bronx, Occupy the Hood (claiming seventeen branches by February 2012), Occupy Boston, Occupy Missoula, Occupy Nashville, Occupy Wichita, Occupy El Paso, Occupy Birmingham, Occupy Tupelo, Occupy Walnut Creek, Occupy Tampa, Occupy Oakland, Occupy Harvard, Occupy Homes, the *Occupied Wall Street Journal,* Occupy the Vote, Occupy Congress, Occupy Everywhere; "Occupy My Heart" (a song): OCCUPY YOUR HEART WITH LOVE and I SHOULD BE GRADING PAPERS BUT RIGHT NOW I'M OCCUPIED (placards at a demonstration). By December 8, 2011,

143 towns and cities in California, nearly 30 percent of the total, had Occupy sites posted on Facebook, one having gone up as early as ten days after Zuccotti Park (Petaluma), followed by South Lake Tahoe and Arcata (September 28), the Coachella Valley (October 2), and Half Moon Bay (October 5). There were, of course, the various slogans and logos pitting 99 percent against 1 percent, like: NINETY-NINE TO ONE: THOSE ARE GREAT ODDS, and hand-printed placards of this sort:

ONE %
?!?!?!?!?!
I BARELY HAVE 1¢

and

I inherited money
at 21. I have had health and
dental insurance all my
life. I want to live
in a world where
we all have enough.
I have more than
enough. Tax me!
Rich kid for redistribution!
I AM THE 1%
I STAND WITH THE 99%

99% could be superimposed or projected, like the Bat-Signal, on the side of a tall building near the Brooklyn Bridge during a November demonstration. By the end of December, a *New York Times* reporter wrote, a Russian dissident on a hunger strike, addressing a rally, was "calling the protesters 'the 99 percent' and saying that Russia was led by a corrupt 1 percent of bureaucrats and oligarchs." If one had waited for such a moment of popular uprising for years, if one had never thought one would live to see such a thing, then it was easy to get swept along in a rapture of self-regard and omnipresence—a euphoria built on the feeling that it was miraculous that such a movement could have come about in the first place, that simply by enduring for several months it was doubly miraculous, that *simply by existing,* and continuing to develop, for all its deficiencies, it had in some crazy sense already won its first battle (as opposed to the contrary long view that expressed itself in the poster, WE ARE THE BEGINNING OF THE BEGINNING). In other words, it was easy for the movement to admire itself in the looking glass, to fall in love with itself.

There emerged, moreover, to identify the movement—brand it, in the current lingo—the home-grown institution of the people's, or human, microphone, which hooked together assemblages into assemblies with its mic check and phrase echoes and twinkling routines. The people's mic was so ingrained in the movement that it was sometimes used as pure ritual, when there was no practical need for amplification—as indoors, or at an otherwise rather cut-and-dried December 1 labor rally in Union Square, where some speakers would speak only a few words at a time into a functioning microphone, allowing the crowd to shout the words back. (And this after a march whose chants were the humdrum "What do we want? Jobs! When do we want them? Now!") The people's mic didn't always work, nor did the rest of the rituals. Some meetings broke down in chaos. There were disputes about how many blocks precisely could obstruct a consensus. Some spoke of a system of "modified consensus." Some considered a GA that employed these techniques "an antidote to capitalist alienation," offering "a fundamental precursor" to real political power, namely "the power to trust again in each other and our collective humanity." The whole ritual lent itself to satire.

Some say that the human amplification system first emerged in the antinuclear protests of the eighties. An-archist organizer Brooke Lehman, who teaches the theories and arts of direct-democracy facilitation, recalls it cropping up during the antiglobalization demonstrations of 1999 in Seattle, when the police denied a crowd of eight hundred, demonstrating in front of the jail, the use of amplification equipment. The tactic didn't catch on, but if one believed that tactics had spirits, one might say that it hovered about waiting to be recalled. The system of hand signals was in use at the 2007 convention of the revived SDS in Detroit; the facilitators there later resurfaced as facilitators in Occupy. In 2011, it was re-launched as a recourse of necessity, the NYPD also having banned electric amplification, but soon enough the human mic became a distinct ritual—a ritual of distinction, in fact—used when the police were in no position to interfere. If the human mic slowed deliberations to the molasses point, so be it—some argued that "well beyond simple amplification, the people's mic allows speakers to know that they are actually being heard," that "by repeating other peoples' words, we are forced to actively engage with them—to actually hear them," that the mic was "an extraordinary tool for opening channels of empathy and solidarity," and that in any event, the slowdown was healthy in an insanely accelerated culture. "In a society where we go apoplectic if we have to sit behind a car at a traffic light or a web page takes more than 2 seconds to load," one Occupy Boston activist wrote, "we must be aware that [we] are not acculturated to have the patience for this process. We're a 'bigger, better, faster' gang."

Some who chafed at the inefficiency were charmed by the calls-and-responses, even touched by their liturgical quality ("almost like a choir, like a modern religious revival," said Jesse LaGreca, who gained renown when a video of him telling off a Fox News reporter went viral). But also there were radicals who considered the mic check a noxious fetish, a turnoff for potentially interested bystanders, an instrument of groupthink, even a Stalinist mind-smothering exercise, and liked to wiggle their fingers derisively, as if putting the twinkle in air quotes. So even as the General Assemblies—almost universally known as GAs—swore by the human mic, there were some who swore *at* it, at its inefficiencies and cultishness and the unfocused quality of the everyone-comes-everyone-participates GA style itself, its way of encouraging the newly radicalized or just plain talkative to jam the gears of a movement that, after all, wanted to change the outside world. "I've been to the GA," said Max Berger, "and it may be worse than Congress."

Berger, twenty-six, with a chiseled face, a trim reddish-brown beard, and a sober manner that gives way to playfulness once you're far into a conversation, is committed to what he calls "the left wing of the possible," precisely the same phrase that, for years, the democratic socialist Michael Harrington used to describe his own political inclination. Berger, for his part, went from anti-Iraq-war activism in high school to Howard Dean's campaign and get-out-the-vote work to the Progressive Change Campaign Committee, J Street, and Rebuild the Dream, enough to be "disillusioned with the institutional left," its stodginess and self-satisfaction. In September, he was feeling "fed up and disgusted, at the end of my rope," and cynical about Occupy's first week until the East 12th Street pepper spraying on September 24 made him think he should check out Zuccotti Park, where one thing led to another, and he was hooked. He was particularly struck by Cornel West telling them, "Don't be afraid to say the word 'revolution.' This is the American Fall, inspired by the Arab Spring.'" Soon Max Berger was a full-time Occupy organizer.

6. The Evolution of Horizontalism

Four months in, the Occupy movement was not con-templating doing away with the people's mic any more than it—if one could even speak of this sprawl of a movement with a singular pronoun—contemplated do-ing away with governance by General Assembly. There would be, however, evolutionary modifications.

A kind of anarchism of direct participation has be-come the reigning spirit of left-wing protest movements in America in the last half century. There is a lineage even longer. Decision-making by consensus is of Quaker in-spiration, as if to say: Speak and listen, listen and speak, until the spirit of the whole emerges. In its recent incar-nation, anarchism is not so much a theory of the absence of government but a mood and a theory and practice of self-organization, or direct democracy, *as* government. The idea is that you do not need institutions because the people, properly assembled, properly deliberating, even in one square block of lower Manhattan, can regulate themselves. Those with the time and patience can frolic and practice direct democracy at the same time—at least until the first frost.

In the house of anarchism are many mansions. The main earlier manifestations, in the late nineteenth and early twentieth centuries, included the strongly individualist style of European (principally Italian) inspiration and the anarcho-syndicalist style of the Industrial Workers of the World—the Wobblies. There were strong anarchist streaks in the New Left of the 1960s, stronger than the socialist streak, in fact, despite all the work Marxists did to conjure proper class categories for the student movement. "Participatory democracy," a phrase that Students for a Democratic Society originally used for a system of self-government both as an end-goal and as a means to achieve it, came to refer primarily to the latter. "Let the people decide," which was one of the early SDS rallying cries, often meant, in practice, "Let's have long meetings where everyone gets to talk." "Freedom is an endless meeting," was a saying of the Student Nonviolent Coordinating Committee. *De facto*, this meant that politics was for people who, in a sense, talked for a living, in other words, college types. It was a revolutionary idea, at least for its time and in certain places: In the Deep

South, for civil rights organizers and black farm workers just to meet and talk was a dangerously radical, and radically dangerous, proposition.

But the left's distrust of outside authority reached, and still reaches, much further. *The* prime bumper sticker of the 1960s New Left might have been Bob Dylan's lyric "Don't follow leaders, watch the parkin' meters," cheekily pairing hierarchy with overregulation. In the dark heart of Mississippi, Bob Moses, one of the civil rights movement's most admired figures, took to standing in the back of the room rather than the front, aiming to depose himself, in a sense. One of the New Left's most accomplished orator-activists, the Berkeley Free Speech Movement's Mario Savio, bridled at being singled out by the media as a leader. By 1967, its membership soaring, SDS was so disdainful of the formal structures of its first five years, so suspicious and resentful of leadership— not just the privileges of leadership but sometimes the hubris of believing that anyone knew more than anyone else, at least about important things—as to abolish its own presidential and vice-presidential offices. The zeitgeist did not take kindly to leaders—at least formal ones.

But, as SDS discovered in 1969, when quarreling Guevarist and Maoist-Stalinist factions tore the organization apart, suspicion of formal procedures didn't keep tiny hierarchies from exercising decisive control at key junctures. They could always claim to be avatars of the Revolution, an abstract enigma that didn't vote them in and couldn't vote them out. Radical feminists savaged their own leaders and were tyrannized by their own structurelessness (to paraphrase the title of a celebrated analysis by feminist political scientist Jo Freeman). Nevertheless, hostility to elitism remained all the rage. From the early seventies on, activists were in revolt against just about anybody's authority, even their own. Vertical authority had a foul odor: It smacked of colonialism, patriarchy, bad white men lording it over voiceless minions. In left-wing activist circles, establishments of all sorts were the immoral equivalents of *The* Establishment.

Accordingly, disgruntled by big-talking leaders, turned off by celebrity-obsessed media, the left of the seventies developed a horizontal style. They accorded limited authority to their own leaders, who were frequently at pains to deny that they were leaders at all. "Affinity groups," small, functional associations of people who knew and trusted each other (a prerequisite for intelligent functioning in chaotic and dangerous circumstances), replaced organized factions and parties. Even movements that seemed to require some level of verticality, those with concrete goals, like banning nuclear power and weapons, or opposing apartheid, were mostly leaderless, at least on the national plane.

Given the growing immigration northward from south of the border, it was only fitting that the horizontality style receive a strong ideological boost from Latin America. The American left heralded the Zapatista movement in Chiapas, Mexico, not only for its embrace of the indigenous, not only for the romance of its masked leader, Subcomandante Marcos, and its hypermodern communication strategies, but for its consensus-based decision-making and autonomist ethos. From the 1990s on, accelerating during the financial crisis of 2001, Argentine radicals and unemployed worker movements committed widespread civil disobedience, leading to antigovernment uprisings, general assemblies, continuing workplace occupations (known as *recoveries*), and the affirmation of *horizontalidad*. In 2010, there were said to be 205 worker-controlled companies in Argentina, employing a total of 9,362 workers. By 2004, the explicit contrast between horizontal and vertical organizational styles had become commonplace at the World Social Forum—the international left's riposte to the annual Davos World Economic Forum—so much so that some radicals warned against "framing . . . this debate in binary terms (vertical versus horizontal)," arguing that it was heightening the risk "that the division could harden and become entrenched, [with] horizontality becoming an identity formation which defines and delimits itself to a specific group of people: 'the horizontals'."

The oppositional culture was shifting, realigning, with the growth of a new marginal style. On the edges of the culture, vertical was stodgy, horizontal was edgy; professional was stodgy, amateur was edgy; orchestrated was stodgy, expressive was edgy; melodic was stodgy, rasping was edgy. From the mid-seventies on in the United States, a political culture of horizontalism intertwined with the defiant, in-your-face style of the punk movement—itself a reaction against the previous countercultural wave, which came to be seen as smiley-face hippies readymade to be co-opted. The rawness of punk, its do-it-yourself, anyone-can-play-music ethos, was coupled with an aesthetic that could also be adopted piecemeal—piercings, tattoos, dyed hair. Defiance met willful amateurism. Traces of this style—and more than traces—would inhabit, occupy, Zuccotti Park.

The afternoon of August 2, 2011, a hundred or more activists, inspired by the June Bloombergville camp, met at the bronze bull in Bowling Green. The gathering had been advertised as a "People's General Assembly on the Budget Cuts" to "Oppose Cutbacks And Austerity of Any Kind" and to plan a September 17 occupation of Wall Street: "Take the bull by the horns," declared the artful poster depicting the head of the frantic, eponymous bull, and adding a definition: "To confront a problem head-on and deal with it openly." In the event, according to Yoni Golijov, a Columbia University undergraduate who took part, the gathering seemed set to devolve into a conventional rally organized by the vertically organized Workers World Party, a tiny but vociferous Marxist-Leninist-Maoist sect that dates from the 1950s (when its founder supported the Soviet suppression of the Hungarian revolution) that came equipped with prefabricated demands like "A massive public-private jobs program!" and "An end to oppression and war!"

A Greek anarchist and avant-garde artist named Georgia Sagri shouted out, "This is not a GA. We need a GA!" She ran into an acquaintance, fifty-year-old anarchist and learned anthropologist David Graeber, and the two of them went around recruiting on the spot, spotting

T-shirts (Zapatistas, Food not Bombs) of likely suspects from "the horizontal crowd." Sagri put up a sign, declaring that an authentic GA would start a few yards away. Twenty or thirty went with them. Then ten others, from the original crowd, convinced them to return to the larger body, where they were able to convince the Workers World faction that they had no support; whereupon the latter departed, perhaps to encounter some actual workers elsewhere. The remnant, numbering about eighty, broke into working groups on food, medicine, Internet, facilitation, media, and the arts, which reported back to the larger group, which in turn passed proposals, including a set of rules to govern subsequent GAs. They proceeded to issue a call for the September 17 occupation.

So, even before settling into Zuccotti Park, Occupy spawned groups galore, as if hell-bent on confirming Alexis de Tocqueville's famous observation about the American social genius for forming voluntary associations. These were *working* groups, functional, skill-based, required, eventually, to meet at least weekly. There were groups for media and outreach and labor outreach and sanitation and the library—complete with actual volunteer librarians—and technical support and medical aid and Safer Spaces (to fight harassment). Security got donations of walkie-talkies. Alternative Banking started, including the former British diplomat and antiwar crusader Carne Ross, along with various experts, ex-administrators, lawyers and bankers, thinking about legislative reforms as well as what a new banking system might look like. (In January, it would rename itself simply: Occupy

Bank.) Medical, using a green tent, included Support, a group of some forty mental-health workers, mostly social workers by day, mostly female, working two-hour shifts all night. (They would produce a manual called *Mindful Occupation: Rising Up Without Burning Out.*) There was an infirmary, filling everybody's prescriptions, helping campers, and at the same time modeling universal health care. There was a Community Alliance, which meant self-policing. There was a life coach, giving free advice on how to plug into Occupy. There were publications galore—the four-page *Occupied Wall Street Journal, IndigNación* (separately edited in Spanish, not a translation), and unaffiliated but friendly adjuncts like *n+1's Occupy! An OWS-Inspired Gazette,* an abstruse and intelligent theoretical journal called *Tidal*, and others.

Working groups, online via the compendious site www.nycga.net/groups, multiplied, leading to subgroup proliferation and some confusion. To a request for clarification of the difference between the Tea & Herbal Medicine group and the Medical Herbalists group, for example, David the Medic, of the latter posted:

> The Tea & Herbal Medicine group actually has always had the dream of becoming more involved in actual herbalist practice, something that we have been helping them with to the extent we can— but they do not actually practice clinical herbalism for the occupation. . . . What we have done throughout the whole occupation, on the other hand, is more like actual clinical work—someone comes to us and

says, "I feel like I'm getting the flu," or, "I've had a UTI for a couple of days," or, "I'm feeling really burned out," etc. We provide a tailored medicinal response to their complaint, plus some general self-care guidance and whatever other assistance we can, and provide continued follow-up care for them as an individual patient.

There were meditation circles—Tibetan Buddhists, for example. And a pop-up Sukkah arrived for observant Jews on the harvest holiday, courtesy of OccupyJudaism NYC and the orthodox Chabad, among other groups. On the eve of Yom Kippur, a thousand Jews gathered in the plaza of Brown Brothers Harriman, across the street from Zuccotti Park, for the Kol Nidre atonement service.

The most conspicuous circles, sonically at least, were the drum circles. The drummers, at the west end of Zuccotti Park, were at it all day and night until neighbors protested; especially in light of the fact that the local Community Board had endorsed Occupy's goals by a large margin, the GA was persuaded to limit drumming to two two-hour stretches during daylight, and then down to one, even after one drummer protested, "We are the movement's heartbeat! You're cutting out your heartbeat!" whereupon somebody else objected, "How is it that one group can claim to be my heartbeat?"

And still, there were solitary drummers who refused to adhere to this community-made law, claiming that all governments, including the GA, were equally oppressive, and unimpressed when accused of usurping power and evaluating themselves into the moral equivalent of predatory speculators contemptuous of all regulation.

The occupants sought refuge from a society of sharp elbows. But—or therefore—they had to cope with people who were far more desperate, or gravitated to public spaces for their own reasons, overlapping but different, like some of the homeless, unsurprisingly, seeking food, warmth, and sometimes drugs as well as companionship, who had sometimes been (or so it was rumored) directed to the parks by the police, and whose experience in decision-making assemblies was decidedly limited by their lower-class backgrounds, and whose resentments of the educated were hard to stifle. After two or three weeks, a good number of the sleepovers were vagabonds, roving from encampment to demonstration like Deadheads roaming the concert circuit. At a reoccupy demonstration on December 17, I met a perfectly decently dressed young woman from Florida, a movement itinerant, who had taken part in a June march at West Virginia's Blair Mountain, protesting plans for mountaintop removal while memorializing the 1921 pro-unionization march on which armed guards had opened fire. At an Occupy Town Square gathering in Washington Square on January 29, a young man from Lawrence, Kansas, carried a sign that read:

Traveling occupir [sic]
needs free hugs
Donations welcome

(He was visiting his fifth encampment, he told me, and donations were arriving.) Some other vagabonds were self-designated crusties, self-consciously filthy, rooted in the punk style and ethos, shoving the label "dirty smelly hippies" back at the Murdoch press and the likes of Newt Gingrich, who had thrown down the gauntlet to all of Occupy: "Go get a job, right after you take a bath."

For all Gingrich's effort to cast himself as a free-spirited rebel during the Republican free-for-all for the presidential nomination, he couldn't resist putting his standard-issue authoritarianism on display. In so doing, he highlighted a theme that authoritarians always throw at anarchists—that they became unruly because their elders, who were supposed to know best, failed to subject them to the heavy upper hand. Spare the rod and spoil the child: This is their shibboleth. When children misbehave, the parents must have been derelict in their duty. When students act up, the faculty must have failed to discipline them.

What Gingrich and his fellow believers were missing, however—what they were constitutionally unable to see—was that during the past decade, in crucial respects, the political classes of major nations in large measure had forfeited their authority. They failed to prevent the attacks of September 11, 2001. They were deceitful and hysterical about Saddam Hussein's danger to the West. They failed to anticipate or prevent the financial crisis, and they failed to provide just remedies for its immense damages. Anthony Barnett summed up the implications for Anglo-American political elites:

> It is not just that political leaders, so-called intelligence communities, and armies with a duty to protect, have together both misled voters *and* proved themselves incompetent. Neither would be a historic first. What is different is that from the start very large sections of we, the people, proved to be wiser than our rulers. We saw further and proved to have better judgment: thus reversing the traditional legitimacy of our elite governance: that those in charge know better than the unwashed.

7. Splendors and Miseries of Structurelessness

After a while, journalists realized that the camps' tribal ways were not an absence of order and structure, but a presence, if a curious one. One feature of their presence was, of course, that they were camps. People slept there, awoke there, and generated electricity with a stationary bicycle there. People were choosing to camp out on a hard rectangle in the shadows of tall buildings. But another, defining, feature was the regular General Assemblies. Everyone could speak at a GA. The principle was equal standing and the norm was generosity. Minutes were posted, and freely commented on, and disputed, and continued online. Intercity cooperation was arranged on free conference calls, sometimes link-ing as many as a hundred far-flung Occupiers, sometimes projected by speakerphone to larger groups. As I correct the copy-edited manuscript of this book, during Occupy's supposed winter lull, an e-mail arrives from an InterOccupy listserv listing *twenty-three* conference calls scheduled over the next five days.

The GAs, and the working groups formed accordingly, were designed to be inclusive. There were complaints that some people there were more equal than other people, that those of privileged background dominated, and that women and people of color were ignored. If women spoke less frequently then men, or more pithily, then periodically the facilitator would pause to see if any women wanted to speak. If persons of color or women (whether "female-bodied" or "female-identified") felt inhibited, or excluded, diversity would be sought, sometimes with a policy called "progressive stack," an affirmative action policy for speakers' lists. If transgender persons resented being called by the wrong pronoun—"he" for a male in transition to female, say—then at the beginning of the next meeting, people would declare not only their first names but the pronouns with which they preferred to be addressed. If the ritual was off-putting to skeptics—talk about political correctness!—they would have to defer; for any discomfort they might feel about this ritual was a price worth paying for conversation in which anyone could feel welcome (unless you felt estranged by the whole process, in which case you refused to pay the psychic cost of admission).

To some of the more experienced and practical-minded activists, concerned less about occupation as a way of life and more about getting decisions made and plans implemented, the rituals could feel cultlike and the commitments required inordinate. Occupiers themselves, busy with meetings elsewhere, stopped attending GAs even when they were making decisions that affected their own lives. "The GA tests the limits of patience for sure—many simple decisions (like how to transport laundry) can take well over an hour," wrote one fervent supporter. Newcomers and bystanders were not always thrilled. Working groups run on the same principle of unrestrained expression could elicit the same exasperation. Another complained to her friends after one particularly contentious People of Color working group meeting: " 'I don't have time for this,' I said, the quintessential New York objection." Since there was no overall budgetary process, lengthy discussions of specific expenditures were slapdash. As in other communities—political, religious, cultural—where the will to belong overpowered the sense of commonality with those who did not belong, these groups drifted toward self-enclosure.

Many were the ways in which the movement could come to feel that its primary achievement was itself—a sort of collective narcissism. However meritorious the rituals of inclusion, an obsession with process—or the wrong kind of process—drove some participants crazy. Then one might well recall that in German, *der Prozess* means "trial," as in Kafka. Larger discussions of tactics, message, and so forth were obstructed by discussions of minutiae, especially financial. Small numbers of boisterous individuals could paralyze the GA. For long periods the GA's online site was obsessed with a proposal to expel a group called The 99 Declaration (the99declaration.org), who supported the creation of a continental congress of elected delegates to be summoned to Philadelphia (where else?), on July 4, 2012 (when else?) to pass a slate of demands to be made on the government. Decisions snarled, as in Congress, and participants snarled. Things came to such a pass that, in December, one participant, Sean McKeown, posted this note: "When the nonmilitant atheist has to guard the door at General Assembly against the virally militant atheists who are trying to disrupt the truly harmless 'grounding meditation' that has been asked for by the vibe checker, it's just another day at the Occupation." To which another, Stuart Leonard, replied: "What happened to the sense of community and respect, or was I just dreaming?" Still, it was a mark of how seriously the ideal of transparency was taken—the General Assembly was the movement's only legitimate leadership, after all—that its proceedings were live streamed and its minutes posted, sometimes with a verbatim transcript.

In response to the widespread feeling that the GA was too unwieldy to make decisions about logistics and direct actions, the Zuccotti Park assemblage decided, in late October, after much debate, to set up a second body, the Spokes Council, made up, physically, of a circle of members of working groups, the other members of which, in principle, would sit behind them, available to caucus when necessary. (*Spokes* referred to wheels as well as authorized spokespersons.) For efficiency's sake, the Spokes Council would meet indoors, three evenings a week, and because it would be compact, it would not require the human mic, though it would retain the sign system: the twinkling, the procedural point triangles, the blocks. The members would be required to rotate weekly, a sort of term-limits system imposed upon a representative body that had been set up reluctantly in the first place. The GAs would go on, discussing general matters, but not the minutiae entailed in, say, financial decisions—no small matter when more than $500,000 had been donated to Occupy and allocations as small as $2000 each had to be decided by consensus.

The Spokes Council began meeting on November 7, but soon became snarled itself. The rule of rotation broke down. The working groups did not seat themselves behind their spokes, and the spokes did not communicate with them. Within a week, even as the Zuccotti Park atmosphere grew nastier—and just before the police shut down the encampment—a woman's caucus "expressed collective outrage at being blatantly disrespected and sidelined" by the Spokes Council. By December, Spokes Council meetings were so chaotic that one activist, Meaghan Linick, likened them to Jerry Springer shows. "There are a lot of angry people," she told me. A few screamers—*paranoids* was another term—were blocking serious proposals. It was "ridiculous," she thought, that a meeting of forty representatives could be paralyzed by as few as three insistent time-hogs. It was undemocratic to debate small matters for three hours as a handful of people succeeded in blocking consensus. The Spokes Council was scarcely immune from the pettiness of the General Assembly.

Brooke Lehman, who had actively promoted the Spokes Council idea, was dismayed. "It was a shock to me to create the Spokes Council," she told me, "and have it be a total shit show." She had been running daily courses on methods of direct democracy, and training facilitators in Zuccotti Park, since just after the occupation began. Her trainees, in turn, had trained others. At the start, most of them were older white men, but now most were very committed young people, under twenty-five. She had spent years doing this sort of training, and was now thrilled to revel in the pleasure of finding herself speaking daily to the newly radicalized who wanted to learn from her. Only now, the Spokes Council meetings that she had spent a month preparing for were so raucous and dysfunctional, the best among her facilitators were heading home depressed.

Lehman thought long and hard about what was going wrong, and came to the conclusion that the central problem was unacknowledged, or half-acknowledged, class differences, which overlapped in turn with race tensions. "Our occupation attracted a lot of people with enormous economic challenges," she said. "Many of them were homeless. In the first few weeks they played a huge role holding the space down." Now, some of them grew resentful. Mistrust festered. After all, they had been living together in one square block for weeks. The homeless and other newcomers grew "suspicious of people who had organizing experience," who were, or seemed, dominant and knowing by virtue of their connections with people who knew people in the organized left and labor, or by virtue of vocabulary and syntax and the rest of what the educated sometimes call *cultural capital*. Even having to form a working group felt like an insult, a hurdle for homeless people looking at a world that they had thought of as their own and was now governed by a structure dominated by affluent white kids. Having half a million dollars or more to allocate (without a budget, yet) "makes it ugly," Lehman said. The homeless, disproportionately people of color, did not feel properly understood or appreciated by "people with the leisure to spend all day occupying," namely people who enjoyed the privilege of "some level of safety net." When people of color spoke out of turn, the reaction was, she thought, "blatantly racist," or at least poor people of color "heard it as racist" and reacted with accusations accordingly. And, at times the reaction was overtly racist, like the sign brought by a white person to a Spokes Council meeting that read: Aryan Brotherhood. The result, she said, was "heartbreak for everyone."

The Occupation was an endless meeting, sometimes a raucous one. There was no provision for child care (not that Zuccotti Park would have been the best place for that under the best circumstances). People with full-time jobs had other things to do than organize OWS. But, it seemed to me, sitting in a dysfunctional Spokes Council meeting the evening of December 29, in a frigid church space—with bad acoustics and facilitators who lacked commanding voices—that class and race tensions were not the only problem. After OWS was evicted from Zuccotti Park, a number of working groups were left without any clear purpose beyond maintaining themselves. Participants who wanted to put specific items on the agenda, like the question of how money was being spent and whether any was missing, felt, to put it mildly, frustrated. One man could take up several minutes during a ten-minute question period without actually asking a question, and when finally challenged, respond, "I don't have a question because I won't talk to authority." A frustrated woman known for regular outbursts lamented: "Let's stop talking about how to get things done, and *get* things done." In a meeting of one hundred, not all of whom had actually been designated as group spokes, wild egos felt free to express themselves without discipline. No matter that many greeted each other warmly at the start; half were so frustrated they left midway through.

"If we're to have a movement, we have to move," another man said more in sorrow than in anger. "We passed this [Spokes Council] process on October 27, it's been two months, and we haven't accomplished a thing."

8. And Leaderlessness

In November, this notice appeared in the fifth issue of the *Occupied Wall Street Journal:*

> Occupy Denver elected the movement's first leader. Her name is Shelby, and she is a 3-year-old border collie dog. Because Shelby can "bleed, breed, and show emotion," Occupy Denver reaffirms that she is "more of a 'person' than a corporation." Occupiers also demanded that "U. S. law be followed as it concerns providing an adequate interpreter," so it's up to the mayor to figure out Shelby's demands.

Here was the spirit of Occupy at its most waggish, as the movement thumbed its collective nose at designated leaders, of course, but also at the clueless journalists and politicians, disingenuous or not, who thought that official leaders clothed in official designations were essential to the grown-up functioning of the civilized world. The movement was prepared to admit that it might be convenient for the journalists if they had leaders to flock to, but to the activists it was a point of pride that they could do just fine without them, and indeed, that the leader fetish was neurotic and laughable, and that, for example, when a committee in Occupy Philadelphia proposed formation of a negotiating committee made up of rotating members from every working group—"a proposal so sound I can barely stand it," wrote an astute labor organizer and Occupy Philadelphia activist—"a sizable portion of the GA sniffs vanguardism, and proposes instead that the city [government leaders] come down to GA—an amendment so insane that I begin to doubt the capacity of my fellow assemblymen and women to govern themselves." Watch the parkin' meters indeed!

But, quite often, as I made my way from one Occupy contact to others, somebody would tell me that I should talk to so-and-so because he or she was "a leader," or "a nonleader," or "sort of a leader," or, after a hesitation, "Well, you know, one of the leaders." Leadership is not abolished when movements don't designate spokespersons and leaders refuse the label, any more than prisons are abolished when they are designated as correctional facilities. In all social groups, leaders emerge. They emerge in the course of action when acts of leadership take place. Leaders prove themselves. Some are labeled leaders, some are not. Some accept the label, others reject it. Those who get a reputation for leadership get treated as leaders. It is as simple (and as complicated) as this: Leaders are persons whom others follow—admire, heed, recognize. Personal qualities count—different ones at different times for different followers. Rhetorical flair counts, again, in different styles for different constituencies. Leaders rise to occasions, size up situations and take a shot at mastering them, and when they make mistakes, prove themselves adept at learning from them. Leaders listen.

The one certainty is that leaders whom no one follows—into action, identification, ideology—are not leaders at all. Self-appointed leaders, even ones who become famous or notorious in the media spotlight, fall of their own weight. The sum total of a process of mutual recognition is what makes leaders. It does not take formal structures—constitutions, officers, titles, elections—to make them. They come into being at moments that call them forth. They act, they are acknowledged and followed; they work with other leaders and go up against them. When movements are organized top–down, the leaders are easy to find. They hold press conferences, give speeches, submit their names and credentials to reporters. When movements fail—or refuse—to fit the established templates, the established templates make them. A movement of anarchist inspiration, like Occupy, does not lack for leadership—it has plenty of leaders, though it rejects the trappings and suits.

And then, new leaders are always emerging, at least to occupy their moments. Thanks to smartphones and the Internet, the torrent of media has swept over the old banks of information and opened new channels to recognition, which is one dimension of leadership. Within Occupy's horizontal world, celebrities emerge from the world of viral Internet videos. That is, they are celebrated by the movement's supporters before they become singled out by official organs of publicity (which sometimes, but not always, catch up). The apparatus of celebrity has need of them, seeks them and finds them. Two who emerged during Occupy's first month were Jesse La-Greca (see chapter 1) and Sergeant Shamar Thomas, a Marine veteran of the Iraq war who was caught on video wearing fatigues and roundly chastising NYPD officers at a Times Square rally on October 15 for throwing their weight around when they confronted nonviolent marchers. "There is no honor in police brutality" was his mantra.

Such individuals might take their turns as Occupy's public faces, but they did not fulfill its internal needs. They could not necessarily steer the movement, give it continuity, help it strategize, devise, and debate rival positions, evaluate successes and failures. There was a graver problem with leaderlessness than the fact that it made it awkward for outsiders to know who to speak with. By rejecting leadership continuity, the movement remained in motion, mobile, able in principle to adapt to new circumstances. But it also rejected the formalities, even the informalities, of accountability. When it made mistakes, it didn't know what to do about them. It was prone, in difficult hours—and all movements, like organizations and marriages, have difficult hours—to thrash around.

Instead of formal leaders, the movement had facilitators, many of them, who met in their own working group. Facilitators kept up conversation, inhibited big talkers and big interruptions, kept conflicts manageable. One member of the OWS facilitation working group observed that because "people tend to conflate visibility with leadership, seeing the facilitators as leaders," it was advisable to train a lot of them, and to make sure they included women and people of color. Facilitators practiced a modest and necessary skill. The problem was, even at their best, they sometimes only lubricated a stalled machine.

9. The Movement as Its Own Demand

On October 5, in Zuccotti Park, I approached a very young man sitting at a table bearing a sign that read "Internet." He seemed, at the moment, unoccupied. I asked him what he wanted to see Occupy Wall Street accomplish in, say, the next six months. He didn't hesitate: "I don't want to have to carry currency issued by the Federal Reserve Bank." Well, I followed up, apart from your own hopes for yourself, what about this movement as a whole? What would you like to see the whole movement doing in six months? "I want to convince other people to have nothing to do with the Federal Reserve System," he repeated.

It was not hard to find "End the Fed" signs—long favored by the libertarian right more than the liberal left—in Zuccotti Park. Nor was it hard to find reform slogans favored by traditional progressives. One favorite, however obscure it might appear, was "Restore Glass-Steagall": the 1933 law that for sixty-five years kept commercial banks from mixing themselves up with securities trading. In 1998, Citicorp and Travelers Insurance had announced their desire to merge. Robert Rubin, who had recently stepped down as secretary of the treasury, brokered a deal that produced the bipartisan Financial Services Modernization Act of 1999, signed by President Clinton, repealing Glass-Steagall and thus permitting investment banks, insurance companies, and other such financial enterprises to merge. Unbeknownst to the public, Rubin was at the moment working out the terms of his next position, as executive without portfolio and director of Citigroup. The repeal of Glass-Steagall was not the whole root of the financial breakdown, but it was historic and exemplary of the breathtaking deregulation onslaught that left pivotal decisions with global import in the hands of venal players.

It was not hard to find signs in Zuccotti Park calling for a tax on Wall Street transactions (sometimes known as a Robin Hood tax), or total public financing of election campaigns, or cuts to military budgets. In October, a Demands Working Group proposed a massive public works program only to have the idea shot down by the General Assembly amid more general mutters against the idea of making demands in the first place. There were calls for

a constitutional amendment that would declare forth-
rightly that the Fourteenth Amendment should not be
construed to give corporations the protections of persons.
(The most pungent statement of that point appeared on

a sign which read: I'LL BELIEVE CORPORATIONS
ARE PEOPLE WHEN TEXAS EXECUTES ONE!)

It ought to have been obvious what the movement
stood for. Anyone with an ear could figure out the essentials. The loudest, most frequently chanted slogans on
the largest marches were "We are the 99 percent!" and
"Banks got bailed out, we got sold out." The first meant:
The plutocracy that controls the commanding heights of

the economy and politics needs to be curbed. The second meant: The federal government under both George W. Bush and Barack Obama caved in to the big banks while failing to relieve householder debt or stop foreclosures. So much was clear to anyone with the curiosity to listen. The general thrust of the movement was unmistakable. It was no more or less vague than the civil rights movement of the early sixties, which after all had called itself "the Freedom Movement," and adopted the call-and-response chant, "What do you want?" "Freedom!" "When do you want it?" "Now!"

However, what was truly impossible to find in the vast reaches of the Occupy movement—for more than three months—was a *single* demand, or a distinct package of them, or, indeed, any specific demands endorsed by the Occupy Wall Street General Assembly or its decision-making equivalents in other cities. Still, even the sternest defender of the politics of demands would have to admit that during the movement's autumn growth spurt, the lack was mainly a strength. True, the absence of demands was taken by mainstream media as a deficiency, a sign of unseriousness. And it wasn't only the media who interpreted demandlessness as a shortfall, for there were many Occupy agnostics who, either innocently or cynically, thought that demands were the measure of a movement's worthiness.

Eventually, the taboo on demands was lifted. On January 3, 2012, the Occupy Wall Street GA, following the example of Occupy Los Angeles, resolved that "Corporations are not people and money is not speech," a measure at the same time being proposed to New York's City Council. The GA also endorsed protest actions to mark the second anniversary of the Supreme Court's decision in Citizens United, in which the Republican-appointed majority held that corporate spending in political campaigns could be limitless. The taboo was weakened. But, at least during the movement's fall-long growth spurt, what Occupy intuitively understood was that steering away from specific demands meant steering away from potentially fierce conflicts over what they ought to be. If the essentials were clear, then demandlessness was tantamount to inclusiveness. Moreover, in keeping with the movement's anarchist, antiauthoritarian thrust, there was a strong sentiment that, as naturalist Gabriel Willow told a *New York Times* reporter, "Demands are disempowering since they require someone else to respond." Demands conferred legitimacy on the authorities. Demandlessness, in other words, was the movement's culture, its identity.

In August, in preparation for the occupation, a red-bearded anarcho-syndicalist from White Plains, David Haack, who was well schooled in anarchist theory and considered both major political parties plutocratic, had urged the GA to make demands, but he too came to think it was a good thing that he had been voted down. The New York General Assembly did recognize a Demands Working Group, but it was preoccupied for months with a jurisdictional dispute. The 99 Declaration group, which wanted a continental congress of elected delegates, and went so far as to run commercials on the Current TV cable channel, and to reserve space in Philadelphia for July 4, was denounced vitriolically on the Demands site as unauthorized and illegitimate.

Meanwhile, on September 29, less than two weeks into the occupation, the New York GA did endorse a "Declaration of the Occupation of New York City," directed, like its 1776 prototype, "to the people of the world," but directed against corporations, not monarchy:

> We write so that all people who feel wronged by the corporate forces of the world can know that we are your allies. . . . We come to you at a time when corporations, which place profit over people, self-interest over justice, and oppression over equality, run our governments. We have peaceably assembled here, as is our right, to let these facts be known.

Then, like the prototype, it went on to itemize grievances, but rather than confine itself to paying "a decent respect to the opinions of mankind," they went further, and urged people everywhere "to assert your power." This declaration received almost no attention in mainstream media. In sum, the journalist Jeff Sharlet formulated Occupy's overall attitude well:

> The question of demands, in all their variety— whether to make them, when to make them, what to demand—is a peculiar one in that it's at the heart of the national occupation debate, and yet mostly irrelevant to the occupiers at Wall Street. Their de-

mand is simply for a better world, which, as far as they're concerned, they've already started building. . . . The divide in the park might be better expressed as between those who didn't believe that the demands group even counted as a part of the occupation, and those willing to let them propose their demands before shooting them down.

10. Wonders of Nonviolence

Sociologist Alex Vitale rightly refers to "Occupy Wall Street's defiant style of nonviolent protest," and it's worth attending to both "defiant" and "style."

The defiance was obviously a matter of attitude—exercising a right, a prerogative to occupy public space, "peaceably to assemble," no matter who says otherwise. Defiance was resistance, reactive; not only did it require an enemy, but it borrowed some of its force from the very force it resisted. For certain, Occupy did not cringe in the face of authorities. It did not feel that they—politicians, police, or anyone else—held any rights superior to their own. Occupy was not given to pleading or lamentation, or, necessarily, what others might call good manners. Even if some protesters drew attention to the insults and injuries meted out to them, they plainly believed in hubris. "Whose streets? Our streets!" This clamor, frequently heard, was one way of putting the claim. But, of course, the chant also demonstrated the limits of defiance, since drivers and nondemonstrating pedestrians might well feel that the streets were equally *theirs*—not just because the law said so, but because citizens were entitled to equal protection. To dictate to rush-hour drivers that their rights should grant priority to those of a protest might well be defensible in a certain version of democratic theory—there is, after all, no constitutional right to convenience—but there was a political price to be paid. In the pure Gandhian version of civil disobedience, it was permissible, even obligatory, to break a law if and only if one were willing to wager one's own freedom, to subject oneself to penalties, to brandish one's own sacrifice in an effort to redeem not only the protestor but the protested-against. After all, to Gandhi, nonviolence was a way of life, a manifestation of *Satyagraha,* "soul force." As Max Berger put it, "You need the right amount of hubris." And that right amount has to be rightly timed and placed.

Occupy's nonviolence was a style, a self-presentation, but it was more than a performance. It was not contrived. Affirmations of nonviolence came up again and again in the Direct Action Working Group (DAWG). Nonviolence in the movement was, said Meaghan Linick, both strategic and moral. It ran deeper, expressing some inner conviction, a sense of dignity that people reached in this movement, some sense that they were, whether taken one at a time or *en masse,* inviolable, that they would only lower themselves if they replied in kind to the violence they received. Perhaps this generosity of spirit was communicated to the uncommitted, or even by opponents, for strikingly, Occupy met with no violence from enraged citizens, or, for that matter, counterprotests. So, in November, the day after Lieutenant John Pike of the University of California, Davis, put a name and a face on police barbarity with the pepper-spraying seen around the world, when chancellor Linda Katehi left a public meeting where she refused to resign and walked to her car, thousands of students sat on the ground lining her route in utter silence—unthreatening, judgmental, bearing witness. Violence made for powerful images, but so did the Davis students sitting. So, while the movement's reputation could be hurt by publicity about so-called violent clashes, it was able, in the main, to claim and hold the high moral ground.

The most graphic exception took place, not surprisingly, in Oakland, where raw police violence has clashed for decades with a ready population of militants, and *agents provocateurs* are hard to distinguish from the provocative home-grown variety who think that escalation is morally righteous and other people's property is an inconvenience if not a theft. I myself witnessed the force of Oakland street disruptions during the Stop the Draft Week demonstrations of 1967, which the organizers called "militant self-defense" and which I thought of at the time as "insurrection." Now, with the extravagant bluster that marks a certain brand of nasty militancy, an anonymous leaflet materialized in early November, signed OAKLAND LIBERATION FRONT, entitled "ARE YOU A PACIFIST?" It opened with these unsubtle lines:

> *YOU* hold the cock of the Empire in your supple hands.
> *YOU* think you are against corporations, the bailouts, the banks . . . but you are actually unknowingly conspiring with them. . . .
> *YOU* are their unpaid soldiers. . . .

The uncomradely meaning was that nonviolence was worse than ineffectual—it was objectively a gift to the status quo, a pillar of false consciousness, shoring up an illegitimate ruling class's claim of the right to impose *its* self-serving conception of violence.

Property damage, in the view of some, did not do any real harm except to the 1 percent and its defenders. On November 2, 2011, amid a much larger rally of peaceable demonstrators in downtown Oakland, several score dressed in black balaclavas (inspired by Europe's violent anarchists, known there for their destructive tactics as "the black bloc") started smashing shop and bank windows as nonviolent activists tried to stop them. Later, the mainstream demonstrators inscribed apologies on the plywood boards that went up over the damage.

The pro-Occupy press expressed dismay at the escalation, whereupon nonviolence prevailed for almost three months.

Eventually, however, the symbolism of property damage proved, to some, irresistible. On January 28, 2012, a thousand or more Occupy Oakland marchers, having been evicted from their outdoor encampment, headed for an unoccupied convention center, intending to claim it as a new headquarters. A few score masked their faces with bandannas. Some marchers tried to tear down the fence around the convention center. Most retreated to their original encampment site, then decided to march through downtown toward a different building, whereupon the police blocked them. Some threw paint cans and bottles at the police, who fired rubber bullets, launched tear gas and flash (or stun) grenades into the crowd. Some three hundred occupiers were arrested (after being told to disperse when the police were spraying them with tear gas and preventing them from dispersing) and, thereafter, as twilight fell, other demonstrators entered City Hall, smashed a glass case containing a model of the county courthouse, and one, wearing a black mask, took the American flag down from the grand staircase, brought it outside, and, as a few in the crowd shouted, "Burn it!" set it afire, as a woman screamed "Stop!" and, according to an AP photographer on the scene, "threw the flag to the ground and tried to put out the fire, shouting at them that it would only hurt their cause." American flags had been burned during the Port of Oakland shutdown in November, but this one was caught by a mainstream photographer, in better light, and in front of City Hall. Meanwhile, the local chapter of the National Lawyers Guild, which frequently posted observers at Occupy demonstrations, received many reports of assaults on protesters, including an incident in which police knocked one person's teeth out with a baton strike to the face. Police reportedly threw others through a glass door, and down a flight of steps. A videographer was pushed to the ground and clubbed.

However, the images that sped around the world were images of a flag-burning. Since September, American flags had become commonplace adornments on encampment tents and at Occupy rallies. Only in Oakland, to date, had any Occupy incendiaries repeated this inflammatory, self-isolating act of the Vietnam War era.

Up to this point, nonviolence had been central and definitional to the movement. It was rational. But, in the course of human history, there is no shortage of social benefits that get passed up or relinquished despite their advantages. Was it strictly on its merits that the prevalence of nonviolence could be explained? I put this question to Matt Smucker, who has been an organizer for more than half his life (seventeen years out of thirty-three), since high school, having grown up in the socially conservative Mennonite church, in Lancaster, Pennsylvania, where he was politicized, he says, by reading the Bible. Afterward, he worked in the Catholic Worker movement, Earth First, the Rainforest Action Network, the anti-globalization movement, the War Resisters

League, and MoveOn. He is a pro—intense, precise, savvy. His disposition can be discerned in the title of the website he helped start, *Beyond the Choir (A Forum for Grassroots Mobilization)*. Nine days into the occupation of Zuccotti Park, he remained profoundly skeptical, publishing a blog post titled, in no uncertain terms, "Occupy Wall Street: Small Convergence of a Radical Fringe." Smucker declared:

> The smallness and fringeness of the Occupy Wall Street protests is symptomatic of a much broader cultural pattern [in which] politics is more about individual self-expression than about strategic engagement. . . . Radicals, like a lot of other people, are caught up in their own self-selecting, self-reinforcing information universes. A few nodes in their network put out a call to occupy Wall Street, all their "friends" repost and retweet, and suddenly it seems that the whole country may just be on the brink of revolution.

A few days later came the seven hundred arrests on Brooklyn Bridge, then the union-aided Foley Square march of October 5. Occupy was no longer small and no longer fringe. Smucker dropped other pursuits, flung himself into Zuccotti Park, and committed himself to Occupy's Public Relations Working Group.

His answer to my question as to why the movement remained nonviolent was elegant: peer pressure. In Seattle, in 1999, he had seen window-smashing anarchists seize the popular imagery by shattering Starbucks and

Nike storefronts, as if multinational corporations were going to be rocked by exhibitions of petty property damage. This time, Smucker thought, the spirit of nonviolence was stronger, embedded more deeply in the movement, tested in what they had all been through together. Transgressions were disapproved of. Social networks had built up. Window smashers, car torchers, and bomb throwers would think twice. Another way of putting the matter came from Max Berger, who also worried about the movement's "solipsism": Any black bloc that thought about seizing the initiative with violence would have to realize that to do so would be tantamount to taking leadership—in a movement that deplores leadership. It would be caught in the jaws of a paradox.

Yet, as Oakland showed, there were other social networks in and around the movement—ones that stood ready to damn the paradox and up the ante. These squads exerted their own peer pressure. Increasingly, in the course of the winter, they found allies with whom to defend what they called *diversity of tactics*, meaning that even some activists who professed that *they* were nonviolent were resolutely unwilling to renounce others in the movement who defended themselves physically or damaged property. ("Why accept the state's definition of violence?" one said in response to my own appeal for absolute nonviolence during a day-long Occupy event in Washington Square the day after the Oakland flag burning.) They valued solidarity over nonviolent principle. Despite the movement's extraordinarily nonviolent opening months—without which it could not have grown to such proportions—it wasn't clear how long it could enfold the two separate camps that crystallized at Occupy meetings: those who would tolerate violence in the ranks on the principle of diversity of tactics, versus those who, recognizing that the violent few almost automatically seize the spotlight and arrogate to themselves the power to shape how the movement looks to the outer world, repudiate them outright. Late the night of January 31, 2012, OccupyWallStNYC tweeted: "We are a non-violent movement. Stand Strong with MLK, Gandhi, Steve Biko & others." Quotes from King and Gandhi sprouted there—signs of commitment, but also anxiety.

Movements learn—fitfully, incompletely, inconsistently, but they learn. Some of the learning is personal. Like Matt Smucker, a fair number of Occupy activists of 2011 were veterans of the Seattle mobilization of 1999 and its successors, and had seen the so-called anti–globalization movement founder on sectarian fights about the merits of violence and disruption, even before the al-Qaeda massacres of September 2001 dropped the movement in its tracks. However, Occupy's social learning wasn't confined to personal experience. Something of a collective lore was being passed down, if incompletely. Even in a country where new generations are declared every few years, and ignorance of the past can be taken as a badge of trendiness, a larger intuition had come to prevail on the left: the simple truth that most of the consequential movements of the second half of the twentieth century were nonviolent movements. They were not tempted by armed struggle.

War was a human tradition, of course, as was guerrilla war, and civil war, but nonviolent struggle was also a tradition. In modern times, it began with Gandhi, and the expulsion of the British Empire from the Indian subcontinent. It continued with the civil rights movement, whose first heralded victory—though not its very first victory—was the Montgomery bus boycott of 1955–56. The student sit-in movement picked up where that left off, in 1960, integrating the South's lunch counters, swimming pools, and other white-only spots, and daring the local vigilantes and racist police to stop them. Defiant nonviolence indeed. Its appeal was electric. On your own initiative, you and your comrades abolished segregation—right here, right now. You did not demand that the authorities end it. You picked yourselves up and ended it, and placed the onus on white supremacists to stop you, and won hearts and minds, and changed the world.

True, as later skeptics would maintain, there was a complication. For many in the movement, nonviolence was less an article of faith than a mystique surrounding something more modest—a strategy. Discussion on this point was not taboo. There were armed black men in the Deep South (Louisiana's Deacons for Defense was the best known of these least known) who had the backs of nonviolent activists against Ku Klux Klan attacks starting in 1964. Many of the Deacons had been in combat, in

World War II and Korea, and especially being Southerners, they were comfortable with firearms. They admired the civil rights movement but thought they needed protection, and local movements were not above accepting their help. Still, as long as nonviolence won victories, most notably the Civil Rights Act of 1964 and the Voting Rights Act of 1965—in no small part by spotlighting racist brutality—it kept up momentum, while liberals could always point offstage to the Nation of Islam and other defiantly non-nonviolent black groups (waiting in the wings, as the expression went) warning against, as James Baldwin wrote prophetically in 1963, "the fire next time."

Nonviolence never went away, but it lost stamina and was outflanked. This was partly because it was far more effective at overcoming legal segregation (whose ugliness was, by 1965, a matter of national consensus) than economic inequality, police brutality, and other racist realities. The social base of the movement shifted from the Deep South to the North and West, where the long-time African-American Christian-pacifist motif of identifying with the so-called suffering servant was weaker. Not long after the triumph of the Selma nonviolent campaign and the Voting Rights Act came the lethal Watts riots. In 1966, Stokely Carmichael's popularization of the studiously ambiguous Black Power slogan released a lot of pent-up rage and inserted a new sound bite into the culture.

Enter the Black Panthers, sexy young black men in leather jackets carrying guns on armed patrols—ingenious investors in the national, then international image bank. From their start, in the San Francisco Bay Area, also in 1966, a mist of romance gathered around them. Masters of dramaturgy, they were lionized on the left for the charisma of their prime leader, Huey Newton, and later the talented writer (and convicted rapist) Eldridge Cleaver, whose compelling essays were published in *Ramparts* magazine. With their bravado came attention, and targeting by various police forces, and gunplay that resulted in numerous killings on both sides. Eventually they set up a Breakfast for Children program, feeding thousands nationwide, for which they were celebrated. In the seventies, they ran for, and sometimes won, local political office. But in truth, neither they nor their imitators could claim many concrete results, and eventually Huey Newton's drug-fueled sadism became well known.

As black militant gunfire died down, and white would-be revolutionaries in the Weather Underground gave up their own bombings and resurfaced, nonviolent civil disobedience continued, developed, became more refined. The movements against nuclear power and nuclear weapons were devotedly nonviolent. By the nineties, Greenpeace and Berkeley's Ruckus Society were rigorously training activists in nonviolent techniques. It did not go without notice that the 1989 uprisings in Eastern Europe were (with the partial exception of Rumania) nonviolent, though it was also evident that the nonviolence of Tiananmen Square ended in a massacre. Still, in no small part because of the wisdom and restraint of Mikhail Gorbachev, hundreds of millions of people lived through a radical transformation of political systems without bloodshed. No wonder that, on the day of his death, Vaclav Havel, the most prominent single individual associated with that monumental achievement, was remembered with a moment of silence at an Occupy conference (see p. 140).

Most recently, and most vividly, the uprisings in Tunisia and Egypt were nonviolent. The overthrow of the ensconced dictators Ben Ali and Mubarak, with relatively few lives lost, was testimony to the practical proceeds as well as the vitality of the Gandhian ideal. How could they not loom large in the imagination of Occupy? By contrast, even the eventual downfall of Libya's Moammar Gaddafi in an *armed* uprising could not detract from the known fact that many thousands of lives were lost in the process. Disciplined nonviolence, if remotely conceivable, was overwhelmingly the right way for a people to begin a new political life while minimizing "collateral damage."

There was even a sort of nonviolent international to train activists, circulate manuals, stoke fires—not that they had any direct influence in Occupy, but they were becoming a force in the international movement of which Occupy was one manifestation. In 2004, veterans of the inventive, successful nonviolent movement that had overthrown Serbia's Slobodan Milosevic founded CANVAS, the Center for Applied NonViolent Action and Strategies. CANVAS learned from the theorizing of the American pacifist Gene Sharp, and published its own sprightly, clever manuals, full of practical tips, like *Nonviolent Struggle—50 Crucial Points* in English, French, and Spanish, and online in Farsi (17,000 of the latter were downloaded during the 2009 Green Movement, which failed, however, to bring down the Ahmadinejad regime after it stole the election). It had good reason to claim influence in the so-called color revolutions of Georgia and Ukraine, as well as the overthrow of a dictator in the Maldives. Egyptian activists of the April 6 movement attended CANVAS workshops in 2009, and went on to play crucial parts in the movement that deposed Mubarak in 2011. Gene Sharp's work had its own devotees in Occupy.

Not that the Egyptian denouement was so happy. A year after Hosni Mubarak was deposed and his party disbanded, the frequently unbridled military continued in power. Islamist groups were bristling with votes. Christian Copts were murdered. All this would make it clear— if it weren't already clear—that nonviolence did not solve every social problem, that Tahrir Square was by no means the entirety of Egypt, that nonviolent liberal activists had little social base in the rest of the country, that the dictator's departure would not transform a corrupt social order, or generate jobs for the millions of young unemployed who had fueled the eighteen-day upheaval that ousted Mubarak. CANVAS itself recognized that the deposition of dictators left ideological and strategic voids, and that nonviolence had to be adapted to strengthen democracy and human rights. Nonviolent movements were not necessarily bound for glory. In 1947, the nonviolent liberation of India from British rule did not avert the savage violence between Hindus and Moslems that resulted in as many as one million casualties. The nonviolent liberation of Eastern Europe from Soviet rule produced autocracies and liberal democracies alike. Of course, there is no reason to think that violent liberation movements would have produced any happier political outcomes—most likely, they would have cost many lives with no better results.

All of which is to say that nonviolence, in both theory and practice, is neither complete nor foolproof. But Occupy's prevailing nonviolence itself is rightly seen by most of the movement as in itself an extraordinary achievement, though one in jeopardy, it would seem increasingly so, from the small minority (overrepresented in New York and Oakland) who contemplate resorting to "black bloc" tactics either as payback against police violence or because they think they can do material damage to the 1 percent's interests. They may be "sweet-tempered" beneath their masks—I heard them described that way by a seasoned anarchist organizer—but they insist on their right to take shelter beneath the movement's "diversity of tactics" umbrella. Still, even as the paramilitarized police provoke reactive rages, nonviolence remains the rule. "The 99 percent is 100 percent nonviolent" is one emergent slogan. This is a point of pride, and surely helps the movement make a favorable impression on the larger public.

The movement's great majority rightly understand nonviolence not as a negation, the absence of destructiveness, but as a creative endeavor, a repertory for invention, an opening, an identity. Occupy does not take nonviolence for granted. It holds workshops—though perhaps not rigorously enough—to train demonstration monitors as to how to contain provocateurs and control large crowds. MoveOn.org and other supportive groups added their own training on a large scale. When theoreticians crop up to argue for a laissez-faire attitude toward tactics, critics step up to dispute the point. Thus, in January 2012, somebody at a meeting of OWS's DAWG exalted the Weathermen (later Weather Underground) for the October 1969 window- and car-smashing mayhem they inflicted upon downtown Chicago under the banner "Days of Rage," generating 287 arrests and an accusation from Chicago Black Panther leader Fred Hampton that the Weathermen were "custeristic." Jim Dingeman, a filmmaker better schooled in political history and old enough to remember the late sixties, promptly replied that the completely nonviolent antiwar moratorium demonstrations that fall were demonstrably far more effective, having helped persuade Richard Nixon to abandon a plan for drastic escalation in Vietnam.

Nonviolence, it would seem, is Occupy's fundamental tradition, not an incidental feature but a force in itself. This simple truth is frequently missed. Theorists of revolution tend to overlook the power of nonviolence as such because they prefer to think about class structures, parties, vanguards, divided elites, and other social forces. This is a retrograde habit. Certainly, the tactics can grow stale with repetition, but committed and creative practitioners can renew it. The Occupy movement has been, so far, a seedbed of creativity. Its future rests in no small part on whether it can continue to learn from experience, deepen its tradition and funnel it into new soil.

11. Radicals

"The core of the group, the people pushing [the Occupy movement] forward," Yotam Marom told me, "are radicals." Meaning what? "They're deep. They're analytical. They go to the root. They're anticapitalist. And they're always nonviolent." He included himself.

Marom's piercing blue eyes and direct gaze do nothing to weaken his definition of *radical*, which ought to be taken exactly as he delivered it and in no other way—however much, in American parlance, *radical* signifies crazy, violent, destructive, as in the phrase *radical Islam*, conjuring up the ghost of Osama bin Laden and tarring generations of Gandhi-inspired organizers with the brush of nihilism. Murderous outbursts are the farthest thing imaginable from what Marom means. Nor does he use the word *radical* merely to provoke. His friend Max Berger uses the word too, and takes pains to clarify: "I'm

both a radical and a reformist." But when Marom uses
the word, he means to speak not only of political objectives but a way of life that endures—a culture that nourishes transformative action over very long hauls.

The notion of radicalism as a way of life came natu-
rally to twenty-five-year-old Marom, the Hoboken-raised
son of Israeli immigrants. After a Labor Zionist upbring-
ing, he attended McGill University, then went to Israel,
where he organized against the West Bank occupation,
and then, with his collective, moved to Brooklyn, lived
in a commune, studied at the New School, and was one

of six activists there who watched an attempted reincarnation of SDS founder there. It was "too structureless," said his colleague, Meaghan Litick, twenty-four, who came to New York from the Detroit area to attend college. The anarchists in the New School group, some of whom "wanted to break windows," she said, were "averse to any sort of decision-making structure," which made it "impossible to get anything done." Marom, Linick, and others declared themselves an organization of organizers, a revolutionary organization involved in mass struggles. They took the name Organization for a Free Society,

OFS had no illusions about achieving solidarity through social networks. They operated face to face. They were serious, smart, and uninterested in recruiting en masse. To the contrary: To join, not only did you have to take part in grass-roots activity, but you had to be vetted in a one-to-one discussion with an old hand, to establish that you shared the group's radical politics. Recruits had to serve a trial period of three months, during which their votes didn't count. Yes, another member, twenty-one-year-old Eliot Tarver, told me, they made decisions with actual votes. "None of *this*," said Tarver, fingers wiggling derisively. Tarver too had moved to New York (from Berkeley) to study at the New School, and got there just in time for the collapse of the new SDS. By December, they had twenty-five or thirty members in New York, some others around Detroit. "We're seen as some of the best organizers—and as the cool kids on the block," said Tarver, and "a place to amp it up to the next level," with analysis and theory. Max Berger was a friend of OFS, but not a member. (Brooke Lehman was a candidate member.) At the same time, there were plenty of Occupy activists who had never heard of the organization.

The OFS group were not, to borrow swear words from the civil rights era, outside agitators. They were inside agitators. They were embedded, said another activist who admired them but called them geeks and overly theoretical. They were neither Marxists, who thought everything in the world came down to class, nor anarchists, who thought everything in the world came down to the badness of state power. They were, said Eliot Tarver, "talking about total revolution—against capitalism, racism, male supremacy." They talked about dual power. "I want a participatory democratic system—in the economy, in the political system," Marom said when I asked him what he wanted to see in five years, adding: "I like to go way out, and think backwards. I want communities governing themselves in assemblies."

Communities governing themselves in assemblies . . . This phrase rings bells. If it sounds like one of the great (and suppressed) ideals in modern radicalism, from the Paris Commune to the early days of the Russian revolution, before the self-governing councils of workers and soldiers were smashed by Lenin's Bolsheviks, and the anarchist assemblies in Barcelona crushed during the Spanish Civil War, it is for good reason. They belonged to the same tradition. If they sounded like the early Students for a Democratic Society—vintage 1963–64, when I was the third president of the organization—they did that, too. SDS was then a face-to-face organization of perhaps twenty chapters and, on paper, some 1500 members, before the Vietnam war brought in thousands in a hurry, many of them more anarchist and rambunctious than the social democratic old guard.

OFS declared: "We understand the problems of our world as interconnected." In 1962, 1963, 1964, SDS's mantra had been, "The issues are interrelated," meaning, more or less, that racial equality required economic reform, and that many of the same figures who sustained white supremacy—especially the Dixiecrats who dominated the committees of Congress—also shored up the military-industrial complex. OFS talked about "dual power" as "a model for getting from the present to the future . . . creating both alternative institutions and counter institutions," "alternative institutions" being "institutions which embody the spirit of our vision for a better world" and "counter institutions both protect[ing] the alternatives and attempt[ing] to convince ever widening circles of people that our holistic vision is, at least mostly, on the mark." SDS had manufactured a button that read: "People should make the decisions that affect their lives," meaning both internally, in the movement itself, and ultimately, in the ideal society.

Marom often wore a scarf rolled into a sort of tube around his neck, over a baggy sweater. Slight and unassuming, he had a quiet look about him, so that it came as some surprise that he spoke so volubly. The popular image of a movement leader tends to define such a personage as a speechmaker, someone who mobilizes masses, but mobilizing is not organizing. Marom was an organizer, "a professional," as he put it, and an organizer is as much a listener as a talker. One of his specialties was shuttling between OWS and sympathetic unions in the process of putting together events like the October 15 march of some twenty thousand on Times Square, which was also a day of smaller, more militant direct actions, for example, a brief occupation of a Citibank branch in Greenwich Village, during which nineteen activists, lecturing the customers about how the bank profited from their indebtedness, and trying to get them to close their accounts, were arrested for trespassing, disorderly conduct, and resisting arrest.

Marom knew that movements to move the larger world need, among other things, strategy. It was he who made the most cogent presentation to a strategic priorities discussion I attended in the Direct Action Working Group, sixty or seventy in number, almost all under thirty, almost all wearing solid colors, the young men slightly bearded, in an unadorned loft near Bowling Green, the Sunday after Thanksgiving. He was respected here; not surprisingly, since the group originated with people he'd worked with on October 15. The meeting was warm, full of hugs, no interruptions. The idea, said the facilitator, who cautioned against anyone taking too long to speak, was "in the course of thirty minutes, to steer ourselves gently toward some shared understanding." But a goodly number of those present did not know the difference between strategy and tactics, so that one person would say "We have to do things even if they're not so exciting," and the next would pop up to say, "We have to occupy something," without giving a reason, and so, jaggedly, the meeting would lurch along, one opinion after another declared but few elaborated or debated, forced to confront other positions. A man in his early thirties addressed the meeting in these terms: "I'm a working dad, I live in Brooklyn, I go home at night, and it's going to be a weird movement if you have to be a homeless kid to participate." He was neither defended nor opposed. His was simply another view.

Marom argued that the movement had to "intervene in social, political, economic processes that actually affect people's lives." That was one crucial means by which it would expand its base. He thought the movement had to map the total situation, and couldn't just give electoral politics the finger; it had to pay attention to what politicians were doing in Washington and Albany. By *political*, a bit shockingly to other activists with purer ideas about stepping aside from traditional government, he meant what most of the country meant: what public officials do and what the public does to change and influence them. Mostly, this meeting consisted of inclusion without engagement, an additive pileup of notions.

"We are not yet a mass movement," somebody said. However, there was no discussion of how to get from the absence of a mass movement to the presence of a mass movement. The radicals were especially aware of the danger of self-enclosure. "Anybody who says there's such a thing as a totally nonhierarchical, agenda-less movement is . . . not stupid, but dangerous, because somebody's got to write the agenda—it doesn't fall out of the sky," Marom told John Heilemann of *New York* magazine. "The GAs," he told me, "empower you, make you feel part of a community, but they have no permanence. People are coming to realize you need an analysis. You can't just go on doing shit. We have to slow down. You can't keep this up. You burn out."

OFS liked to think long term. The OFS website declared that the group was "struggling for a world of equity, solidarity, diversity, self-management, and ecological balance." They intended no false modesty:

> We are committed to building a movement for social liberation. We aim to transform the governing values and institutions in all spheres of social life. . . . We work to break down all systems of inequality and injustice and to create a participatory, democratic, and egalitarian society.

OFS had no qualms about calling itself revolutionary: "We recognize the need to fundamentally transform the governing values and institutions of society." As precedents and exemplars, they gathered a hodgepodge they called "a long line of revolutionaries," roping in anarchists Emma Goldman and Peter Kropotkin, nonviolent heroes Fannie Lou Hamer, Gandhi, Martin Luther King, and SNCC jostling alongside the cruel and paranoid Black Panther Huey Newton, in surely the most inclusive sentence in the history of revolutionary manifestoes: "From the Industrial Workers of the World to the Spanish anarchists, from SNCC to the Black Panthers, the Cuban revolution to the Vietnamese resistance, from Gandhi to King, from the kibbutzim. . . . to the Bolivarian Revolution," thus associating Israel's mostly socialist founders with the Hugo Chavez who once blithely designated "the descendants of those who crucified Christ" as being among those who "took possession of the riches of the world." They were against all forms of oppression and did not think capitalism or racism or sexism was more important than the other two. They thought the financial meltdown was not an exceptional interruption to the guiding genius of the invisible hand, but a predictable convulsion in a global economic system where markets regularly fly off-balance, and the advantages that accrue to making and breaking the rules are so immense, they smother any niggling attempts to regulate it into milder waters.

The radicalism of the core movement helped explain what baffled so many observers—the absence of formal demands and programs. As a guiding principle, what the radicals wanted was direct democracy. It would have been absurd to demand that the authorities create direct democracy. The authorities had everything at stake in resisting such a demand. If you were going to have direct democracy, you had to launch it yourselves, directly. You had to infuse the spirit of do-it-yourself with world-changing zeal, and vice versa. Political-economic decisions were too consequential to be made by anyone but all the persons concerned—the stakeholders, to use the current lingo. The radical core wanted a world run not by exclusive committees but by assemblies of the people.

They wanted neighborhood assemblies and workplace assemblies. And, suddenly, because Occupy Wall Street had created so many new facts, had so transformed the popular discourse, the radicals had gained—however informally, however limited—a hearing. Then there were others in the encampments, probably a majority, who wanted change but had not thought through how it might come; who were not so interested in ideology, had not devoted much time or study to the shape of a new social order, but believed in their own ways that the whole political economy was implicated in the financial meltdown, and were therefore partial to some idea, however inchoate, of reconstructing the whole system. So, at least for the moment, the hopes of the radicals dovetailed with a larger state of mind.

12. The Co-optation Phobia

Sunday, December 19, was a big day for Occupy Wall Street conferences. At the New School in Greenwich Village, Occupy intellectuals put on a full afternoon of brisk, sharp panel discussions about banks, foreclosures, debt, and strategy, in front of an audience numbering a hundred or more. Not far away, at Pace University, near the Brooklyn Bridge, several hundred met for an all-day "unconference," which had the imprimatur of the GA. This was a jamboree of workshops, covering vision, strategy, and tactics—under the rubric "Where Do We Go From Here?"

The largest session, "Visioning," took up the entire gym. The floor was open for anyone to say whatever they liked. There was a moment of silence for Vaclav Havel, that paragon of nonviolent liberation, whose death had just been announced. Facilitators read out various propositions, inviting the ring of two hundred or so present to declare themselves bodily. "If you want Occupy to stay out of politics," he said, "step forward." Almost everyone did. A few minutes later, the format changed. This time, if you agreed with a proposition, you were to walk to one side of the gym; if you disagreed, to the other side; if you weren't sure, to the center. One proposition read: "Achieving reforms is important." Ninety percent moved to the "agree" side.

The Occupy movement wanted to win reforms and to stay out of politics. At the same time.

Movements are social organisms, living phenomena that breathe in and adapt to their environments, not objects frozen into their categories while taxonomists poke and prod them. They come, go, mutate, expand, contract, rest, split, stagnate, ally, cast off outworn tissue, decay, regenerate, go round in circles, are always accused of being co-opted and selling out, and are often declared dead. If they are large, they contain multitudes, and contradict themselves. Outsider movements struggle to finesse their tensions, to square circles, striving to hold onto their outsider status while also producing results.

Can Occupy's tensions be finessed, its circles squared? It partly depends, of course, on the strengths of the arguments pro and con (which I shall try to assess in part III). But it's also important to take seriously the resistance points, the points at which Occupy not only rejects arguments but deflects them. There is fervor in its resistance and it is the fervor as well as the arguments—the fervor beyond and beneath the logic and evidence—that deserve attention. Why the largest and fastest-growing movement of the left in many decades should have declined (with some exceptions) to throw its weight directly into the political campaigns of 2012 is a matter of culture and identity. Occupy does not want to be mainstream. It is, at its core, an outsider movement, deeply committed to a radical departure from political norms. That is its identity, an identity only reinforced by its early flush of success. And such success imposes burdens.

Success? Is it reasonable to speak of success when the plutocracy prevails, when big money still dominates official politics, when the investment banks and their executives thrive with impunity under minimal regulation, when corporate power still rules markets and melts icecaps? Despite a world of change it has not achieved, the movement can still take a certain success to heart—can *feel* success—even if, at some level, it still disbelieves what it has wrought. It burst out of nowhere. Its interior bonds, many of them, are intense. Enough of its inner life satisfies enough of its inner core. Arrests, and the insults and injuries meted out by the police and their ideological cheering squads, consolidate bonds. If the working groups and decision-making structures are only intermittently functional, they have created a sort of way of life. However outlandish that way of life may look to traditionalist outsiders, outlandishness is—to the core— proof that they are authentically resistant.

It is indeed a strange confidence that this movement feels, for it has the strength of a certain contradiction. On the one hand, it thrives on the esprit of the few: "We few, we happy few, we band of brothers"—and sisters. On the other, it embodies the passions, often buried ones, of most of its fellow citizens. It has not only won headlines, and boasted millions of online hits and other signs of conventional repute, but has welcomed an incoming stream of fresh participants, put recruits to work, made use of their skills, whether in designing posters or organizing donated books into libraries or generating electricity, and even having been displaced from its encampments and forced indoors during the winter it continues to improvise communities and to spin off activities that succeeded in being fun and consciousness raising and practical all at once, and even effective at, say, coming to the aid of the foreclosed or convincing depositors to shut down their accounts at the most pernicious banks. But still, high repute counts as a sort of success—how could it not?

Moreover, the movement was fortunate in its enemies. When journalists and pundits, and not a few traditional leftists, were early on taken aback by the roughness of this slouching beast, their scorn and incredulity actually raised the movement's morale. When they highlighted the riff-raff they encountered on the sidewalks, in the tents and under the tarps, paying special attention to the unkempt and, if female, topless; when they derided the welter of slogans and bludgeoned the encampments for incoherence—they had no leaders! no demands! too many demands!—and when a top Republican saw "growing mobs," this was vindication.

So, it was not bad news for the movement when, for example, the creator of the comic book *Batman: The Dark Knight Returns* blog-ranted:

> "Occupy" is nothing but a pack of louts, thieves, and rapists, an unruly mob, fed by Woodstock-era nostalgia and putrid false righteousness. . . . Wake up, pond scum . . .

They were, he warned,

> iPhone iPad wielding spoiled brats who should stop getting in the way of working people and find jobs for themselves. . . . Maybe, between bouts of self-pity and all the other tasty tidbits of narcissism you've been served up in your sheltered, comfy little worlds, you've heard terms like al-Qaeda and Islamicism. [These true enemies] must be getting a dark chuckle, if not an outright horselaugh - out of your vain, childish, self-destructive spectacle. In the name of decency, go home to your parents, you losers. Go back to your mommas' basements and play with your Lords Of Warcraft.

Ann Coulter, promoting her book, *Demonic: How the Liberal Mob Is Endangering America,* the most recent in her series of subtly titled tales about, well, the demonic left, spoke nostalgically of the Kent State killings of

1970. The hysterical right-wing agitator David Horowitz, ever scouring the horizon for scare stories with which to rile his supporters into coughing up donations, e-mailed: "This dark carnival of rape, cop-hating, dope dealing, and even murder has been endorsed by the Democratic Party establishment and even the President himself, because it supports their class warfare agenda."

The crudity of antagonists could be counted, then, as further success for the movement. But success posed problems, too. It was one thing to win the support of the rapper Kanye West, of Michael Moore and Cornel West and other left-wing celebrities. It was another to win over David Letterman ("I love these people causin' trouble"). Cultural success required navigational skills. Just a few weeks into the encampments—the mere flicker of a moment in subjective time—"Occupy" was already a touchstone. "Ninety-nine percent" was the most fashionable label and newest thing. "One percent" jokes ran rife through the Internet. Popular culture coated Occupy with an aura of clean fun, trendiness, and inevitability, inspiring that mixture of disgust, smirks, and sneaking satisfaction which is characteristic of the cultural confusion that thrives when depthless rebellion has become routine, and a proudly, even aggressively marginal culture struggles to remain marginal because—paradoxical as it may sound—the feeling of marginality binds their community, confers a sense of election.

Success meant quandaries. Mainstream culture was always lurching off in search of bandwagons to jump

aboard, so how could you hold on to an outsider identity when Occupy was the latest, hottest ticket?
However urgent the movement's moral passion and
revolutionary desire, however fervently it hoped against
hope that some day the populace might rise against
the depredations of plutocratic control, it had begun
with the default assumption that the public was quies-
cent, neutered, and that significant community was to
be found only on its edges. To find oneself then com-
mended by David Letterman was, in a certain way, problematic. Letterman himself, no fool, got the point, joking
that he would like to have gone downtown to see the
protests, "But I know I would be beaten. They don't like
the idea that famous people with dough are sucking up
to them."

Indeed, establishment figures of any and all kinds
made Occupy uneasy. When, early in October, I asked
some New York militants about Elizabeth Warren, just
ramping up her populist campaign for the Democratic
Senate nomination in Massachusetts, one of them ex-
ploded in disbelief that I could for a moment think that
she might be Occupy's sort of politician, for had she not,
in a Democratic primary debate, told the movement to
obey the law, and had she not been milder in her support
than her rivals for the nomination? (The first charge was
accurate—hardly surprising from a law professor—but
the second was false.) Former House Speaker Nancy Pelosi won little if any praise from movement insiders when
she supported them, though at times, sensitive to fluctuations in the esteem in which they were held by public

opinion, she did tiptoe gingerly around the name, pre-
ferring the locution "grassroots citizen movements working to hold special interests accountable." When she
told Democrats, "We've got to mobilize the 99 percent,"
those who feared co-optation above all else smelled
danger. Democratic icons from the revered (Rep. John
Lewis of Atlanta) to the corrupt (Rep. Charles Rangel of
Manhattan) were shooed away from general assemblies
when they attempted to speak without waiting to take
their turns in the circle of speakers, the point being that
elected representatives deserved no special treatment.
Never mind that John Lewis was one of the grand expo-
nents of nonviolence in American history; he was now a
government official, not entitled to the slightest defer-
ence, and so denied permission to "jump the stack" and
speak without a lengthy wait. It was as if Occupy had
been reading Walt Whitman, twinkling to signal agreement that the great city was indeed a place "where the
populace rise at once against the never-ending audacity
of elected persons."

Part of what the movement ignited, after all—rhetorical triumphs, celebrity endorsements, all the suspect rewards that flow from success—blew back into their eyes
like smoke. The more rewarding it was to find oneself
in a community of mutuality, reciprocity, and warmth,
the greater the fear that it would dissolve. Cachet was
suspect. Even outside support—like tents and medical supplies and portable toilets donated by MoveOn,
which was anti-hierarchical in its own way, and by labor
unions—was suspect.

MoveOn, for its part, was content to operate behind the scenes and did not go out of its way to advertise its support. In the view of its executive director, thirty-eight-year-old Justin Ruben, Occupy was "articulating a critique that's deeper than MoveOn's. One of the things that Occupy is and will be, or will birth, is a voice that's to the left of MoveOn. I want Occupy to say stuff that we wouldn't. Those things can change the whole conversation." MoveOn itself, with its six million members, would be part of a "broader economic populist movement . . . a 99% movement." Many of their members joined Occupy, especially outside the biggest cities, Ruben said, and many were "looking on with interest and excitement." Still, when they spoke of Occupy, they referred to them as "they."

Meanwhile, some Occupiers spoke of MoveOn as a sinister force, the way social democrats once fretted about the many-tentacled Communist Party. Sometimes they expressed suspicion of the MoveOn leadership in particular, though Ruben and others insisted that MoveOn only did what its members decided to do. On one Occupy listserv, a man denounced OWS's habit of "tiptoeing around organizations like MoveOn who have proven themselves to be willing enablers of the system we are now in revolt against," and such talk was not uncommon. In fact, many activists were already suspicious of MoveOn even before September 17; it was, they thought, "not radical or progressive enough," said Occupy public relations specialist Ed Needham. They saw it as a Trojan horse, a "Democratic Party front group," as journalist and Occupy organizer David DeGraw put it, so much so that somebody posts as "MoveOnSucks" on the New York GA website. After all, Needham added, "This is a movement in which a lot of people say the difference between the Republicans and the Democrats is that the Democrats take a little longer to get down on their knees for money." DeGraw said, "We tried desperately to get [MoveOn's] help before 9/17. There's a grudge to be gotten over there. We've had run-ins with them since then. They were having people go on TV, talking as if they were the movement. The biggest fear among organizers that I speak to is, we don't want to become the progressive version of the Tea Party. We want to keep nonpartisan." It was hard to find hard-core Occupiers who disagreed. Even organizers like Max Berger who helped broker arrangements between Occupy and MoveOn shared the harder-line movement's suspicion.

From a certain point of view, MoveOn's aid threatened the integrity of the horizontally organized community. That kind of success was corrosive. Recognition was a certificate of legitimacy from authorities who did not deserve their authority. The movement thrived on a sense of beautiful marginality, but rapture was attached to a sense of vulnerability, for it was good to be supported but bad to be trendy. It was good to be sought after, bad to be captive. It was good to be a community banded together in warm solidarity, bad to be smothered by hypocrites. There were Occupy activists who ceased wanting to hang out with non-Occupy people.

So, no sooner had the endorsements begun arriving

than the encampments were stalked by the bogeyman of co-optation. Heated arguments burst out in Occupy's spaces, online and at meetings. When union people attended GAs, were they threatening co-optation? (On the other hand, if they stayed away, were they outside exploiters?) The radical organizer Van Jones had been tossed out of a low-ranking White House position in 2009 when attacked by Republicans, but when, independent of Occupy (starting earlier, in fact), he tried to put together what he called an American Dream Movement to promote progressive reforms, some Occupy activists scorned him as if he were a White House mole or an agent of a foreign power (if those were not two different names for the same thing).

When politicians made friendly noises, and celebrities recorded songs celebrating Occupy and reproducing some of its images, was this co-optation? A cynical—or sociologically structural—view was that behind the warnings against cooperation stood a fear that somewhere, someone prominent or powerful (and weren't they the same thing?) was having too good a time riding the new tide. Underneath all the political arguments, the fear of co-optation was a fear that the beautiful new thing under the sun might be fragile after all.

"Everyone sees a piece in this movement and asks, how can I get my slice? No. Bake your own pie," Amin Husain told me. "This is fundamentally different from how capitalism thinks."

A soft-spoken but burningly intense and eloquent Midwesterner from a poor Palestinian-American family, he spent years living in Ramallah, on the West Bank, where he became involved in the First Intifada and saw the inside of Israeli jails. He studied at Cambridge and Columbia Law School. Soon enough, he had graduated from dumpster-diving (during law school) to earning $185,000 a year, plus bonus, handling financial transactions and private equity (doing midsize deals, he told me, rattling off the figures "250 million to 1 billion") for the major corporate law firm Cravath, Swaine & Moore. This wasn't the life he wanted. He quit. The next thing he knew, he was living in Budapest, listening to acid jazz, writing on napkins. Refashioning himself as a video and performance artist back in New York, during the Arab spring, he and Brooklyn artist friends put out "communiqués," leaflets drawn from Middle East newspapers with the headlines translated into English. He joined Occupy from day one.

What, I asked him, should the movement do during the election year? Nothing in particular, he said. "I feel like we're losing when we're putting our principles up for sale. To me, that sounds like politicians." This was a pervasive opinion even among occupiers who had worked in Democratic Party politics, like Haywood Carey, who slept in Zuccotti Park for a month and argued, according to reporter Justin Elliott, "that the movement is too young to begin considering questions about electoral politics. 'There is an inherent distrust and disdain for politicians here,' says Carey. 'We're going to continue doing exactly what we're doing.'"

Amin Husain insisted that the movement's nature

was "plural." For him, this was more than a concession, it was a principle. He warned against folding the whole movement into an electoral campaign, or, for that matter, a campaign against foreclosures. "Everyone is jumping in and wants a piece of this [movement]," he told John Heilemann. "The problem is that you start taking what is potentially a transformative movement and start making it into a corporation . . . retarded and nonfunctional."

The movement, Husain thought, was built on "something innate in us that rises up." However, the problem of how such a movement should deal with social realities that are not innate, conventional structures like political parties, had a long lineage. And, in the sixties, the problem of steering clear of co-optation got more complicated, for the upheavals were intimately connected to tidal shifts in popular culture—which is to say, among other things, commercial culture. The decade of sex and drugs and rock 'n' roll was also the decade of the Monkees, the Dodge Rebellion, and Columbia Records' full-page underground press ads of 1968 proclaiming, "The man can't bust our music." Popular culture—and the consumer goods that conveyed it—ceased to market uniformity and quiescence, and instead promoted uniqueness and uprising. (Was it because everything was suddenly labeled *unique*—a devaluating of the currency of uniqueness—that we began to hear about items being *very unique*, uniqueness itself having been downgraded from absolute to relative status?) As the counterculture went over the counter, rebellion became the mainstream of popular culture.

A culture of rebellion was, to put it mildly, problematic. Rebellion through style, after all, is not so much a direction as a stance, not a place to stand so much as a statement about where not to stand. Since popular culture was always hospitable to the new—it aspired to be, above all else, popular—Occupy's incursions into the mainstream sparked anxiety. Commerce works by scenting social movements, softening their edges, and thereby remembering how to stay popular. This is not the rebellion of Albert Camus, wherein the rebel stands "inflexible and free." What Occupy feared was the politically empty shape-shifting of, say, Madonna. Thus, when Rocawear, the urban clothing line co-founded by rapper-impresario Jay-Z, put "Occupy All Streets" T-shirts on sale, and admitted that they would not be sharing profits with the movement, Occupy supporters denounced the company for rank exploitation. Priscilla Grim, co-editor of the *Occupied Wall Street Journal* and a public relations professional with years of nonprofit experience and a flair for trash talk, was quoted on the entertainment website TMZ saying:

> *Jay-Z, as talented as he is, has the political sensibility of a hood rat and is a scrotum. To attempt to profit off of the first important social movement of 50 years with an overpriced piece of cotton is an insult to the fight for economic civil rights known as #occupywallstreet.*

A few days later, the shirts were withdrawn.

Priscilla Grim knows how to convert cultural anxiety to public relations opportunities. She scorches earth. Under her *nom de guerre* Grim Womyn, she appeared three times on Sean Hannity's radio show and, when he hazed her with (true) reports of sexual harassment in Zuccotti Park, zapped him back by congratulating him for his belated discovery that violence against women is widespread. Each of her appearances, she says, brings her new Twitter followers. All well and good—she wants to get the word out beyond the automatic supporters. New Yorkers, she says, know about the Occupy movement, but chatting with strangers at roadside stops on a fall car trip back to her home town in Tennessee, she learned that other folks didn't. What to do about that? "You have to engage with mainstream media," she says firmly. "You have to know how to walk that line."

Late in November, the teen idol Miley Cyrus released a remix of a recent song of hers, "Liberty Walk," this time "dedicated to the thousands of people who are standing up for what they believe in," replete with images of demonstrators marching and being tear-gassed, from New York to Greece and Spain, and inspired characteristic ambivalence and another saucy retort from Priscilla Grim. Grim, a thirty-seven-year-old single mother whose black hair is streaked with purple, who has lived near the financial margin for much of her life, once published fanzines, and helps edit the elegant wearethe99percent Tumblr website, was on the job—the unpaid job she holds as one of the Occupy PR affinity group while also maintaining her own Tumblr, grimwomyn ("all things grim, womyn, and revolutionary . . . or so I try"). She stepped up and told TMZ that while the remix "rocks in spirit, I double dog dare [her] to fight on the front line of economic civil rights at LA City Hall"—adding, "Revolutionaries occupy, Ms. Cyrus." (There were rumors that Ms. Cyrus did want to show up at Occupy LA, but that her "people" vetoed the idea.) The spectacle of a gossip site quoting a mouthy political radical against a music celebrity does raise a pot–kettle problem but underscores the ease with which the boundaries between popularity and commercial success blur in a spongy culture.

The common though unstated assumption in Occupy was that the movement was weak while mainstream groups were strong. This was, said Eliot Tarver, "ridiculous." The movement was not helpless. As Occupy's Matt Smucker wrote:

> The fact that establishment Dems are clamoring to figure out how to co-opt this energy is a serious victory for genuine progressives and Left radicals. This is what *political leverage* looks like. Radicals haven't had it in this country for a very long time, and now we're getting a taste of it.

Others among the political radicals agreed that the danger of co-optation was overstated, and that the key was to distinguish between the Occupy movement, which was not in danger of doing anyone else's bidding, and the larger 99 percent movement which had in it many mansions. "I don't think it's possible to co-opt this thing," Yotam Marom told John Heilemann. In his view, it cost OWS nothing when politicians talked them up. "OWS is the beating heart of the 99 percent movement," he told me. "If other people want a piece of it, great. If Howard Dean wants to endorse the 99 percent idea, fine. But he shouldn't take aerospace money. And if they want to put OWS on the logo for his yard posters, we have a problem with it."

I asked Marom how he would react if unions, MoveOn, and other reform groups came forth to do what Occupy had steadfastly refused to do—make specific proposals. "If they want to make demands, let them," he told me. As for the classical reform–revolution conundrum, he told Heilemann, "I don't think it's an either or. People who only want reforms are probably just handicapped by cynicism. And if you don't want reforms as a revolutionary, then you're not a revolutionary, because people need the foundations on top of which to survive. And people need to win things, to feel like it's possible to win."

Further he would not go. As for him, he would never *ask* President Obama to do something. Making demands on the authorities would inevitably stamp them with approval, would confer on them what, in the prevailing Occupy view, they did not deserve—authority. He was not terribly interested in what the Democrats were up to. In early December, he had not thought much about what would happen during the coming election year. Still, it seemed clear that those most confident in the Occupy movement—those with their eyes on the long-term prize, a major transformation in American life—had the least to fear from the co-optation bugaboo. The only thing they had to fear, it seemed, was fear itself.

PART THREE
THE PROMISE

Shane Patrick, who'd done extended time in the thick-ets of New York City's left-wing politics (he'd first been galvanized by protests against racist police attacks during the Giuliani years) and learned not to get carried away by street actions, visited Zuccotti Park around Hallow-een and felt its intoxication. He came from the punk and hardcore music scene—there, he'd learned to book shows and publicize bands before getting a degree in political sci-ence and history at CCNY (on a student loan)—and was now making a bare living with temp jobs. He volunteered for Occupy's press group. Soon, he was one of a couple of dozen people who answered reporters' queries e-mailed to press@occupywallst.org, publicized the doings of OWS groups, focused on "messaging."

In early January, I asked Patrick what he thought he would be doing in five years. "Hopefully, I'll be employed," he shot back. When I asked him what he saw Occupy doing, he replied, "I don't see it existing in five years." What did he mean? "From outside," he clarified, "the movement seems highly capable. Inside, you're more aware of the intense fragility." "Perpetual whirlpool," was one phrase he used. "Such a swirling thing." Contradictions pulled at it. The allure was romantic, he granted. However, "There's a perpetual sense that chaos is always possible around the corner. It informs the process within Occupy. People coalesced around an initial vision, but it's a huge coalition. If the structures aren't agile and capable enough to accommodate this huge span of people, this coalition could wither and collapse under intense pressure."

The pressure was on. As the election year progressed, would activists who abstain from electoral politics on principle—some even refusing to meet politicians of any stripe, for meeting them would be tantamount to acknowledging their legitimacy—be able to live with the ones who took the electoral calendar seriously or even wanted to work for candidates? Would the anarcho-syndicalists get along with the anti-austerity reformers? How much of Occupy's oxygen would be taken up by black bloc tactics? There were contradictions galore.

Shane Patrick is one of the many skilled operatives who occupy key points in Occupy's expanding universe. Such people exist in any live social movement—they are sometimes mistaken for leaders, though some are relatively visible and others much less so, but there is a remarkable concentration of them in Occupy Wall

Street. There are the carpenters and electricians who set up Liberty Square's power generation (by stationary bike); there are the computer wonks, lawyers, chefs, accountants, doctors, sanitation experts, tech people, life coaches, videographers, and journalists. Plainly, there are elective affinities at work. Personal talents dovetail with the needs of the movement, and they select each other. In network theory, Patrick is a node. For more than a decade, he watched little groups crystallize and dissolve, ebb and flow, in that intense world located within a mile on either side of the Brooklyn Bridge, where post-punk, anarchist, and left-wing grouplets intersect, tangle, and bleed together, endlessly morph into new configurations—and usually stay penned within their enclaves. In 2011, the nexus came out into the larger city, and world, and he plunged into it.

The movement, he thought, had to evolve, or mutate, beyond its time-specific origins. Occupy Wall Street was suited—in many ways brilliantly, for all its disorganization—to a particular moment when several currents met. Anger at the plutocracy was vast but largely unexpressed. Avoidable economic suffering was widespread. The glow of Barack Obama's rise to the White House had worn off. Ed Needham, another Occupy press liaison figure (at "the ripe age of forty-four") who had taken part in the Draft Obama movement—and himself ran for local office three times as a Democrat in Maine—expressed an almost universal sentiment among activists when he told me he was disillusioned within Obama's first hundred days in office. "I'm pragmatic but

I faulted him for not building a popular consensus, capitalizing on his victory. We thought it was the dawn of a new day; it was really just the dusk of the former day." Meanwhile, in the eyes of a good number of activists like Shane Patrick, the Obama glow had never materialized in the first place. They could not be disillusioned because they had never been illusioned, had never, in the regnant cliché, drunk the Kool-Aid.

Whether they knew it or not, critical masses, grousing about everything that had not changed with Obama in the White House and the Democrats in charge of Congress, were waiting for a signal to wake up and try something new. The signal came, the moment arrived, the movement erupted. Five years down the road, as Shane Patrick pointed out, conditions would be different. The economic crisis might even settle into some sort of tolerable equilibrium—all things are possible. Each vector—the thousands of mostly young divergent activists of the inner Occupy movement, who began by reveling in what they could do with public spaces and dispersed after eviction to devise continuations; the outer movement of unions, churches, and liberal groups, itself not durably organized; the Democrats; the Republicans; and not least, paramilitarized police forces loaded with fancy equipment purchased in the name of Homeland Security—all these were vectors reacting to other vectors within a total ecology of social, political, and economic conditions, not only nationally but globally. But it can surely be said that simply to roll back the economic (meaning human) damage done by the global meltdown of 2006-08, let alone to set the nation and the world on a substantially more decent course, would take years, if for no other reason than that the damage, and the abyss of inequality, have grown over decades. A movement to repair economic inequality, let alone reconstruct the economy on more democratic principles, is not going to become superfluous—will not be worked out of a job—in the foreseeable future.

Consider that, in 2008, among the Organization for Economic Co-operation and Development's member countries, the United States ranked higher in income inequality than any country in Europe. The gap between the wealthiest and the rest began to increase in the late seventies and deepened in the late eighties. Between 1984 and 2008, the share of America's total income (including capital gains and dividends) that went to the top-earning 1 percent of households rose by almost 40 percent. In the United States, the share of the top 0.1% in total pretax income quadrupled during the thirty years up to 2008. Meanwhile, the tax rates paid by the richest declined. If income were really a reward for hard work, then the superrich must have been beating themselves into the ground mercilessly, while the 99 percent were lollygagging around their swimming pools.

And, if tax cuts actually did redound to the benefit of the great majority, through growth and trickle-down income, as the Republicans maintain, the country should have been brimming with growth during these decades, and the trickle should have been strong enough to bathe the 99 percent in a great flood of prosperity and upward mobility. Instead, as regulation shriveled, and the entire go-go banking sector went gaga, economic growth ground to a standstill, or reversed, and most of the public lost ground. The rich soaked the less rich. Moreover, upward mobility—a technical name for the American dream, the deferral of progress into the future that is supposed to be our historical compensation for more inequality in the present—was now lower than in Canada and most of Europe. Whatever the folklore said, Americans were not moving on up.

All this took place during a period when the leading figures in not only finance but government, and in the university circles that recruited for them, congratulated themselves that everything economic was coming up roses. This was the time when Alan Greenspan, who chaired the Federal Reserve for the two decades from 1987 to 2006, could be counted upon to deliver himself of such statements as this:

> Those of us who support market capitalism in its more competitive forms might argue that unfettered markets create a degree of wealth that fosters a more civilized existence. I have always found that insight compelling.

At least until (so Greenspan testified after retiring) he was "shocked" to discover that there was "a flaw in the model . . . that defines how the world works," namely, in the belief that he had held to unswervingly for forty years: that banks, operating rationally in their own behalf, would act in such a way as to protect themselves. Calamity was required before scales fell from the revered elder statesman's eyes. And yet, even after Greenspan fell from grace, the variant models on which his successors rely are still, and again, held to be flawless—or flawless enough (for all but the unlucky), or, in any event, the best that semidemocratic capitalism can do.

So, if economic life is to be made substantially fairer and more decent, and plutocratic power is to be reversed, an enduring movement is essential. Such a movement may not be sufficient—it isn't humanly possible to know that—but surely it is necessary. Occupy's thrust is popular, which is essential, but popularity itself does not change the world.

What does? In the longer run, both institutional change and changes of heart and mind. The movement needs structures that flex and learn, train leaders, generate actions, recruit supporters. It needs to be a full-service movement—one that invites participation at many levels. For overmortgaged and underwater homeowners, it needs campaigns to corral the banks that have them in hock. For the civilly disobedient young, it needs appealing direct actions. For those too busy or politically disinclined to do anything but sign petitions, it needs petitions. Whoever you are, it needs to promote activities tailored to you.

In the medium run, say five years, networks of activists—the inner movement—need to devise an infrastructure that sustains them, recruits them, focuses their intellectual and strategic life, generates sustained pressure on power, keeps movement tensions manageable, and not least, make significant progress toward driving money out of the political system. During the election year of 2012, the Occupy movement needs to deepen and widen. It needs to win concrete victories, engage the political system, continue shaping the debate. The 99 percent movement—including the left's membership organizations—need to formulate their own demands and coordinate strategies. The rest of this book details the challenges these projects meet, and evaluates their prospects.

And because the power of money must drastically shrink before we achieve anything like full-blooded democracy—if only because the citizenry need a certain platform of security if they are to invest time in politics, as well as a modicum of faith that democratic action might pay off—the impulse that guides an equality movement is, and must remain, more than an envy-fueled resentment: It must be a love of democratic potential as well. If your goal is even more ambitious than reducing inequality—if you favor greater economic democracy as well, like the Occupy radicals—you have all the more reason to need an enduring movement.

After all, the plutocratic and inegalitarian trends are not going to reverse because the market on its own pulls itself together to roll them back. If the last decades have demonstrated anything—indeed, if the history of American capitalism demonstrates anything—it is that the social arrangement known by the God-term *the market* is perfectly content with vast inequalities. As for the political class, its reliance on big money inhibits, if it does not outright extinguish, whatever reform impulses well up from time to time. Given the power of money in politics, the political class is too interlocked with lobbyists, Wall Street, and the rest of the corporate galaxy to care enough to take the political risks. Democratic vitality is both the prerequisite and the outcome of a continuing mobilization to make the conditions of life more decent and fair.

On the face of it, then, an equality movement or, technically, a more-equality movement, is needed and will continue to be needed. But its form may change, and probably should evolve over time, social forms being no more sacred than species (if you accept the idea that species evolve, which of course many of my fellow citizens do not). Occupy may need to divide, or develop a division of labor—may need to give rise to distinct tendencies or even organizations trying distinct approaches, taking on distinct foci, heading toward distinct aims—compatible at best, but still distinct. But whether, over the medium or long haul, Occupy continues to regard itself as a unity or not, it (or its offshoots) will have to ask, and re-ask, the question of what it needs to be a movement facing outward, a movement that moves the 99 percent, more than a welcoming campground.

13. Live-in Victories

Self-government can be rewarding. Even long meetings can be bracing, if only for the multiple joys of self-expression, especially for people who are not used to anyone (at least outside their immediate circles) paying attention to them. But most Americans are not enamored of political activity of any sort. Insofar as they have leisure time, they can think of plenty of things that might be more fun, or comforting, or comfortable than sitting for hours in vigorous conversation about fund allocations or whether a working group is legitimate. To become involved, they need incentives. Their time is a resource and you don't have to be a Wall Streeter who thinks that nothing motivates people besides self-interest to think that people are inclined to invest their time rationally, or at least to make the effort. The core of the Occupy movement finds incentives galore in the spirit of the movement itself. However, most of America doesn't get close enough, in the first place, to be enticed. Or they get close enough to be repelled by what looks like a disconcerting style. There will always be free riders, waiting and seeing, but if they are to get involved, most people need more incentives than the pleasure of their fellow seekers' company alone: They need to believe that activity brings results. "To keep growing," somebody said at the Direct Action Working Group (DAWG) meeting January 8, "we need regular victories."

However disaffected they are from the everyday political process, movement activists too think that reforms matter, even when the prospect of changing laws is meager, as when the Republicans control at least one branch of government. (I say this not out of any sentimental attachment to the Democratic Party, but because the Republicans are *almost completely* beholden to the plutocracy, while the Democrats have split loyalties—partly to the plutocracy, partly to the majority of the populace.) Even laws when they are passed need execution that is not always forthcoming; results, when they do arrive, are slow. People want results they can see and touch.

From its start, Occupy has been busy, one might even say frantic, performing mostly symbolic actions that impress the activists as "beautiful," "inspiring," "awesome." "We've done a lot!" said one young man at the DAWG meeting, contemplating an imposing list of actions that the movement had undertaken during its first sixteen weeks, adding, "It's kind of cool, but it's also kind of tiring." The movement, on the surface, is a festival of performances. At Lincoln Center, on December 2, when protesters greeted the audience leaving the Gandhi opera *Satyagraha*, and the composer Philip Glass went outside to read them, human-mic speech style, a passage from the libretto, somebody shouted out: "This is the real opera!" Some participants indeed resort to the language of the theater: One DAWG speaker celebrated what he called, "breaking the fourth wall," as on the day when downtown taxi drivers told demonstrators where the police were gathering. The turn of the new year continued in this vein—an "Occupy Octopus" float carried in the Rose Bowl parade, a flash mob in Grand Central Terminal to protest military spending. Creative reinvention is as good for a movement as it is for life in general. When fresh stunts are pulled off—not just repeated—they're good for morale. Social movements also want to have fun, even if the triumphs are fleeting and the costs (in bail money, for example) considerable. Repeat protests, however, run athwart the media's been-there-done-that syndrome, and the movement's own tedium threshold. A demonstration outside Mayor Michael Bloomberg's townhouse on January 6, 2012, protesting arrests of journalists, drew only fifty demonstrators, far fewer than showed up for an earlier such event; press coverage was minuscule. Occupy Congress, on January 17, 2012, a day announced for rallies and lobbying was a bust, with only a few hundred present until late at night, suggesting that standard-issue Washington rallies no longer galvanize. The principle, says thirty-year-old Will Jesse, an OWS publicist who in his day job (which suffered considerably

during peak fall Occupy events) consults on ecosustain-
ability campaigns, is to keep stirring the pot, and to bring
out large numbers when you can. "Keep the people in
power uncomfortable," a fancy, mainstream PR person
advises him, "You guys don't understand how powerful
you are. Just by being annoying."

Annoying the powers that be isn't everything—in
fact, however much it may gratify the inner core, it's of
little value to outsiders, especially doubters who need
evidence that a movement can be effective. But a world
that is dominated by the top 1 percent and their institutions offers a target-rich environment.

On the potential effectiveness front, the most promising
line of action that Occupy has opened up directly addresses the ruination left by the bursting of the housing
bubble. The idea is to stop evictions and foreclosure auc-
tions, and to occupy houses whose occupants have been
evicted, or are being foreclosed upon, or stand empty but
are capable of being made habitable. After all, there are
many more peopleless homes than homeless people in
America, and banks continue to get away with accounting barbarities; for if their assets were valued at current
market values, rather than inflated numbers, they would
be no more solvent than many homeowners discovered
they were after the bubble burst. During the first nine
months of 2011, American banks launched 1.8 million
foreclosure actions, though this number was down from
the record 2.9 million during the whole year 2010. Millions more foreclosures are expected in 2012. Mean-

while, the much-touted agreement between forty-nine state attorneys general and five huge banks, announced in February 2012 as a breakthrough that pried $25 billion loose, netted a meager *two thousand dollars apiece* for the victims of abusive foreclosures. The notion that housing is a human right may seem outlandish in a culture of savage capitalism, but amid the rubble of middle-class dreams, it has bite. It was for good reason that the New York General Assembly's Declaration of the Occupation of New York listed first among its bill of grievances against corporations: "They have taken our houses through an illegal foreclosure process. . . ."

A movement to keep or put people in homes offers a profusion of options. It not only aims to make banks accountable (both for the financial collapse and for ongoing gouging), "It helps specific people," as Occupy Our Homes organizer Matt Browner Hamlin says, although the negotiations with banks that result from occupations produce "more delays than mortgage modifications." While the autonomous Occupy Our Homes, like Occupy, refrains from making concrete demands on the authorities, putting a family in an affordable home delivers a "tangible" result—the word chosen by the Occupy Wall Street housing activist Eliot Tarver. The actions can be readily explained. They federate a host of organizations, some national, some local, that agree on the tactic of occupation. They excite many of the original Occupiers, for physical places get occupied and the little encampments, at least in theory, provide journalists with something to look at. If they require justification, that can be forthcoming. For example, some will object that residents who overborrowed—who should have known that they were not going to be able to make mortgage payments, who should have overcome their own hunger for excess square footage—deserve neither sympathy nor homes. But it can be readily pointed out that fraudulent or near-fraudulent mortgages, aka "innovative products," were not incidental crimes. They were what one financial writer, writing when the bubble was still expanding, called the "driving force" of the housing boom. There were so-called NINJA loans ("No income, no job, or assets"). There were *balloon mortgages*, where borrowers paid only interest for ten years, until large lump-sum payments fell due. There were *liar loans*, in which borrowers stated their annual incomes without any need for documentation. There were loans with zero down payments, and *teaser loans* that started out with very low interest rates, to be reset in two years; and there were other creative wrinkles as well. In Oakland, some were called *firma, fecha* loans—Spanish for "signature, date." Vast numbers of these deals—a majority, by some accounts—were laden with fake signatures and robo-signed by mortgage sellers, including large banks, who collected their fees for making loans unbothered by any need for due diligence. Such frauds loom large to explain how millions of homes became vacant.

A movement of repossession, deforeclosure, and housing occupation could, in today's jargon, continue to scale—mushroom. Such actions already have spread. In 2010, a north Minneapolis homeowner whose house was repossessed and sold by a bank without her knowledge, found organized volunteers to camp out on her property to prevent eviction by Freddie Mac. In November 2011, Occupy Minneapolis joined in and Freddie Mac postponed the eviction. In March 2011, a two-week eviction blockade in Rochester, NY, by "Take Back the Land," failed to prevent a twenty-car police armada from evicting a woman named Catherine Lennon. But in August, she won a court order blocking the eviction. Also in August, a month before Zuccotti Park became world-renowned, a group called Organizing for Occupation had stopped the eviction of an eighty-two-year-old woman in the Bedford-Stuyvesant area of Brooklyn. She had been victimized by a predatory lender and her fraudulent mortgage then traded around banks for a decade. "We're not slaves anymore," the tearful homeowner told the crowd. "My grandfather was a slave, but I'm not." Then, in October 2011, a dozen OWS and other activists were arrested in a Brooklyn courthouse for singing to disrupt a foreclosure auction. "Stop lootin', start prosecutin'," read one sign. On January 26, still another Brooklyn auction was stopped by a hundred more singing activists, of whom thirty-seven were arrested.

Such ideas were infectious. For years, national networks of housing activists had clamored for affordable housing. They were in touch with each other. Communities of sympathetic neighbors developed. Now came the contagion. A one-hundred-three-year-old Atlanta woman, Vinia Hall, along with her eighty-three-year-old daughter, was served with an eviction notice. A local store manager who had lost his own house to foreclosure a year earlier started a campaign to keep them in their house. As promised, or threatened, sheriff's deputies and a crew of movers went there on November 30, 2011, but, in the language of local reporters, decided they couldn't go through with the eviction. In the light of local media coverage, JPMorgan Chase, which held the mortgage, decided to work out a settlement enabling Vinia Hall and her daughter to stay.

Meanwhile, in nearby Riverdale, Georgia, Brigitte Walker—a career army veteran with a 90 percent disability from a spinal injury incurred in Iraq, suffering also from posttraumatic stress disorder, unable to work, her income more than halved—read about Vinia Hall's victory. Over a period of two years, she had tried five times to get JPMorgan Chase to modify her loan, and five times she had failed. Two weeks before her home was to be sold at the Fulton County Courthouse, she contacted Occupy Atlanta. On December 6, a date that had been set forth as a national day of coordinated foreclosure-related actions, she welcomed Occupy activists to her home. Press conferences followed, along with a national campaign to call the bank to protest. Chase agreed to strike a deal.

That same day, in a collaboration called Organizing for Occupation (O4O), combining several housing groups and community activists, who went around fastening yellow OCCUPY tape and FORECLOSE ON BANKS NOT ON PEOPLE signs on abandoned houses, a thousand OWS supporters marched through the East New

York section of Brooklyn, one of the most foreclosed-upon areas of the state, to a two-story house at 702 Vermont St. (Said one experienced housing organizer, Rob Robinson of Take Back the Land, "The Occupy kids were able to bring masses of people together," while "community groups . . . couldn't get people out there. They were beat up by apathy and a lot of other things.") This house, which had been converted to small apartments by its owner, Wise Ahadzi, a former day trader, was taken over in 2008 by the Bank of America (which acquired the mortgage when it acquired the rest of Countrywide Financial, one of the most abusive subprime lenders). The landlord, underwater, having in 2007 spent $424,500 on a building currently valued at $262,300, abandoned the building, which was then looted, turned into a drug den, its yard piled with waste. It still stood vacant. On December 6, Alfredo Carrasquillo, a homeless housing organizer himself, and his wife Tasha Glasgow, moved in. "Deforeclosure" they called it. During the nights that followed, Occupy volunteers slept over, and skilled volunteers began rehabbing the building. Police put in an appearance, but since the landlord, Ahadzi, had not decided to evict them, they left. In fact, behind the scenes, Ahadzi was negotiating with the activists for a change in the bank's terms, so that he would be relieved of his mortgage, the building would be turned over to a community group, and, instead of looking like a slumlord, he would look like a hero. The rehab work went on, and the house on Vermont Street became a display case and rallying point for the O4O sub-movement. But, within a few weeks, the Bank of America offered Ahadzi new and favorable terms. He decided to take his building back. The unhappy occupiers left, after restoring a wall they had torn down. Rupert Murdoch's *New York Post* gloated.

December 6 saw more than fifty occupations across the country, most of them more successful. Some were undertaken by local groups like those in Minneapolis and Georgia that had already led on the issue. Some took place where the national organizers had no contacts at all, for example, Boise, Idaho. The word was spread by anti-foreclosure bloggers and movement videos showing occupations and the rehab process. "This is how you take a home that's been vacant and make it habitable," Matt Browner Hamlin says. This twenty-nine-year-old activist is a veteran of Democratic campaigns, the SEIU, and various nonprofits: a prime example of making a career amid the infrastructure that nourishes the Occupy movement. However shaky the labor movement when it comes to protecting its members and recruiting new ones—immense legal and administrative hurdles stand in the way of success, even though most workers say they would like to join unions—organized labor remains a major training ground for organizers, who carry their skills elsewhere.

Those skills are always in shorter-than-needed supply. A movement dealing with property questions obviously needs to do its homework scrupulously, but does not always do so. In December, an Occupy Oakland encampment settled onto a vacant plot they thought was owned by the city, but it turned out to be privately owned. When a man claiming to be the owner showed up, the police cleared the site at his behest.

The housing focus is popular with the movement, but the question remains, in the words of one person at New York DAWG's strategy session, January 8: "Actions to stop evictions affect one family, but how can we do things like that that affect lots of people?" More coordinated days of house takeovers are in the works as I write. Networks of volunteers who will show up to stop evictions are being organized. There are proposals to focus on particular banks whose practices are especially egregious, like JPMorgan Chase in Atlanta, US Bank in Minnesota, and possibly even Bank of America nationwide, which, having bought all of Countrywide's subprime paper, is particularly overextended—some would say effectively bankrupt already. Indeed, on January 8, a young man who works in finance arose at the DA meeting to say, "I propose that we make the Bank of America fail"—a line that drew appreciative laughter—"which is not an impractical idea, since right now it's a cesspool of legal liability. I think we can make the Bank of America fail in six months." He proposed mortgage payment strikes, debt strikes, a campaign to spread the word at bank branches, ATMs, and shareholder meetings—all actions designed "to let people know about B of A's illegal actions." Many people, he said, did not realize they could challenge foreclosures, that their legal positions were strong because the banks were getting away with using bad paper. Many who heard him were intrigued, and not simply by the two different cloth gloves he wore, one striped brown, the other, red, white, and blue. He was by no means the only finance professional who came to Occupy meetings to offer expertise and advice.

This is not a movement that thinks small. Nor does it think in lockstep, as should be abundantly clear by now. About housing actions there are skeptics inside the Occupy movement. OWS direct-action regular Amin Husain saw early on how hard it was to choose the right houses and the new inhabitants. He saw a lack of transparency in some decisions. He thought the leaders were "privileged people, the kind who get airplay on CNN and MSNBC. They need to check themselves. I want to see the white males step back. They need to listen to the marginalized people. Otherwise you're just a politician."

It wasn't just the process that bothered him. On the merits, he was against folding the movement down into something manageable.

> If we become an anti-foreclosure movement, it's too narrow. We put people in harm's way. Then what happens? When do they move out? We act like saviors, but I don't want to see another Habitat for fucking Humanity. That exists. Charity never achieved justice.

But just as Husain heard his voice growing angry, he reeled himself back: "You can't have one solution for 99 percent of the population. The movement isn't stuck in discord. It's only if an action endangers the whole movement that it's a problem." I asked for examples. "I could give hypotheticals," he said, "but it's not happening in this moment."

Could the movement's overall justice camp coexist with its social reform camp? "We have to be able to," Husain says. "Everyone has to be less ambitious, less certain they're right. We are going to be what we want to see. That was the spirit that was powerful in Liberty Square during the first weeks." Is this utopian? "This is not about utopia, but about empowering. It's like Tahrir Square saying: 'We will govern ourselves.' Imagination is what wins. You have to release people's potential."

But how? And are all releases equally valuable? Could house occupations compete for headlines with tumultuous street actions? The same DAWG participant who at a January 8 strategy meeting touted the need for regular victories gave three examples of what he had in mind: re-occupations of homes, debt strikes, and a general strike May 1. (The idea of a debt strike against student loans has also been floated, but is slow to pick up support.) Someone who had taken part in Occupy Oakland's port shutdown promoted "a national strike against capitalism." Someone else said: "I'm totally in love with a general strike. To me it's like seeing the face of God." The idea of an entire people rising up, putting a stop to all business as usual, remaking everything, launching Year Zero, best of all on the left's traditional day of celebration: This is the luminous dream of the left, and has been so for more than a century. Not surprisingly, it generates more excitement in New York and the San Francisco Bay Area than elsewhere in the country. The speaker did not pause to note that general strikes, to be serious, have to be organized by the workers who walk out of their jobs, not by outside radicals who may well be perceived as privileged agitators parachuting in from outside, and that there is little sign that workers in large numbers are prepared to walk off their jobs. A failed general strike would play in the media as a grand defeat. Nonetheless, over the winter, DAWG support for a May 1 general strike grew.

14. Work the System? Change It? Smash It?

Ideally, says Will Jesse, the movement needs actions that can be replicated anywhere, like the original occupation, which was, as he puts it, an "open-source" operation. The message it sent out was: "We are standing up to Wall Street over here at Zuccotti Park. You can do this yourself." The December 6 housing occupations were other such examples, he points out: "It's so clear that people everywhere were duped into predatory loans, and there are homeless people who can be matched up with unoccupied homes. . . . You want people who hear about this action to say, 'Yeah, we could do that too.'" December 6 was, as somebody said at the DAWG meeting, a media slam-dunk.

Will Jesse's biggest worry is that the movement might fracture over how to relate to the 2012 elections—that members of Occupy's very broad coalition, ranging from reformers to anarchists, from nose-holding Obama voters to a small fraction of third-party enthusiasts, would not only diverge but fight bitterly over whether they even belong to the same movement. Amid the radicalization of movement recruits, some, especially outside New York City, want to work the levers of normal politics, to elect officials and push them to make reforms. Some want to change the political system through constitutional amendments or direct action, or a combination. Some, probably no more than one in ten—though again they are more conspicuous in New York—want to disrupt corporate business as usual and work toward creating assemblies that can sweep aside conventional political structures and, on their own, rule. Anger at potential co-optation continues.

These are built-in disputes, reliably and inevitably erupting during the entire history of outsider politics. Whenever awakenings erupt in American history, promising vast cultural as well as political changes, there are those in the movement who fear that narrowing their focus to trifling results will condemn them to irrelevancy. Many abolitionists feared getting lost inside the new antislavery Republican Party after 1854. Populists feared getting lost inside the newly populized Democratic Party that nominated William Jennings Bryan for president in 1896. In the early twentieth century, the anarcho-syndicalists of the Industrial Workers of the World feared that the labor movement would be usurped by electoral politics. Comparable ruptures divided the civil rights and antiwar movements. Such disputes are intrinsic because movements and parties differ fundamentally. To put it crudely, movements are energy and parties are mass—at best engines, at worst tombstones. Energy and engines need each other but operate under divergent principles. Movements tend toward horizontality and parties run hierarchically. Careers and rewards in parties are drastically different from careers and rewards in movements. Movements must confront marginality but parties must confront corruption. Movements attract the unruly, parties the manipulative, and grand awakenings some of the most fervent and impractical of all. This is not necessarily because any of the protagonists are wicked. But these two very different social phenomena do tend to attract different types of individuals and then sharpen the differences between them.

If the divergent tendencies are to hold together in fruitful rather than destructive tension, despite the centrifugal forces, there must be actions that all can agree on and participate in, actions that look like winners for all, that speak clearly to the 99 percent about the movement's direction, that declare that it stands beside the supermajority against the warping of American life by big money.

In the meantime, little imagination is required to conjure up scenarios for fracture or worse, though several interlocked things have to be said at the outset. Politics does not permit the certainties of, say, the schedule for sunrise and high tides. The theory of social movements is—and must necessarily be—underdeveloped. Prophecy is a fool's game. It crumples when there is a profusion of moving parts. So many vectors collide with, impinge upon, hook up with, and otherwise interact with the others. A force field so thick with vectors does not produce predictable billiard-ball effects. Obviously, a host of moving parts and vectors are in play within, around, and against the Occupy movement, in the total field of action that constitutes it, both feeding it and in the end (if there is an end) limiting it.

There is not only the inner movement, which is, as Will Jesse says, a coalition unto itself, but there is also the outer movement made up of labor unions, reform groups, progressive Democrats, clergy, and others who don't necessarily get along. There is, not least, Wall Street and its allies, their treasuries, organization, and initiative. There are economic and political contingencies of many kinds: Republican and Democratic moves, international factors, and so on. These are not perfunctory disclaimers. They are sober disclaimers.

All this said, the most plausible scenarios for the movement's fracture or worse revolve around events of 2012: one that is certain to take place (the election campaign) and others that are likely (outbreaks of violence).

Simply by existing and commanding the spotlight for several months, Occupy became, and is likely to remain, an issue in both national and state campaigns. Republicans demonize the movement, identify it with radical demands, and try to pin both onto the tails of the Democrats. Not surprisingly, national Republicans targeted Elizabeth Warren as the poster lady of Occupy. On November 9, an attack ad backed by Karl Rove featured this voiceover:

> *Fourteen million Americans out of work, but instead of focusing on jobs, Elizabeth Warren sides with extreme left protest. At Occupy Wall Street, protestors attack police, do drugs, and trash public parks. They support radical redistribution of wealth and violence. But Warren boasts, "I created much of the intellectual foundation for what they do. I support what they do." Intellectual foundation for what? We need jobs, not intellectual theories and radical protest.*

As a woman fervently narrated this thirty-second capsule crammed with attack triggers, the words "Socialist" and "Professor" flared up on the screen.

Warren and other Democrats recognized the line they had to walk: supporting the movement's thrust though not its every tactic or costume. By November, she had worked out a politically astute way to talk about the movement. In a TV interview, Warren replied to Rove:

> *Occupy Wall Street is an independent movement, organic, moving in its own direction. And that's how it should be. It's its own voice. What I'm doing is what*

I've been doing for a long time, and that is protesting the activities on Wall Street, trying to hold those guys accountable, and now trying to take that fight to the United States Senate.

Occupy would run on its own track, she would run on hers, and they would head in the same direction. While most Occupy activists are scornful of Democrats in general, some volunteer to work for Warren in Massachusetts, even as others with equal fervor denounce her as an agent of co-optation. As the time draws nigh for the election, she will surely win the votes of the great majority of Occupy supporters.

Meanwhile, influential Republican pollster Frank Luntz told Republican governors, "I'm so scared of this anti-Wall Street effort. I'm frightened to death. [The Occupy movement is] having an impact on what the American people think of capitalism." He proceeded to counsel his audience with warm fuzzy words to defend Republican policies and blame economic troubles on President Obama. But, even if Luntz was only feigning fear for rhetorical purposes, his advice presupposed that Occupy was moving public opinion. Contrary to Karl Rove, he did not advise a heads-on attack against the movement. He was, in a Republican sense, taking the high road, leaving Rove and the rest of his gutter gang to manage the smears.

As for President Barack Obama, he walks Elizabeth Warren's tightrope but carries much more baggage. Before the occupation of Zuccotti Park, he was performing a rhetorical balancing act, insisting—even if tactically—on postpartisanship in his Labor Day speech in Detroit: "I still believe both parties can work together to solve our problems. . . . We're going to see if congressional Republicans will put country before party." Then, two weeks into the occupation, on October 5, White House Chief of Staff William Daley was asked whether Occupy demonstrations were helpful to the White House's push for a new jobs bill. The former investment banker replied tunelessly: "I don't know if it's helpful." The next day, perhaps as a small corrective, Obama ratcheted up his rhetoric a bit, responding to a question about Occupy at a press conference by saying that "the protestors are giving voice to a more broad-based frustration about how our financial system works." On that occasion, he was reasonably enough accused, by one left-wing blogger, of "tiptoeing" around the question "without explicitly supporting or distancing himself from the movement." There may well have been a debate on in the White House.

But by November 22, when "mic-checked" at a New Hampshire event, Obama let the chanters have their say and then adroitly addressed, and heralded, them as if applying his faith in postpartisanship to the problem of escorting protesters and conventional Democrats into the same big tent. "Mic check!" the Occupiers shouted:

> Mr. President, over 4,000 peaceful protesters have been arrested. While bankers continue to destroy the American economy. You must stop the assault

on our First Amendment rights. Your silence sends a message that police brutality is acceptable. Banks got bailed out. We got sold out.

They were drowned out by Obama supporters, whom Obama quieted with a softly stated "It's all right." He then directly addressed the chanters:

Listen, I'm going to be talking about a whole range of things today, and I appreciate you guys making your point. Let me go ahead and make mine, all right? And I'll listen to you. You listen to me.

He went on:

A lot of the folks who've been down in New York and all across the country in the Occupy movement, there is a profound sense of frustration, a profound sense of frustration about the fact that the essence of the American Dream . . . feels like it's slipping away. . . . [F]amilies like yours, young people like the ones here today—including the ones who were just chanting at me—you're the reason that I ran for office in the first place.

Occupy's Shane Patrick, no Obama enthusiast, called the president's maneuvers on this occasion brilliant.

By December, Obama was deeper into a populist turn by denouncing "the breathtaking greed of a few," warning against "the gaping inequality [which] gives lie to the promise that's at the very heart of America: that this is a place where you can make it if you try," and putting his own spin on the Occupy vocabulary by deploring the fact that in recent decades "the average income of the top 1 percent has gone up by more than 250 percent to $1.2 million per year. . . . For the top one hundredth of 1 percent, the average income is now $27 million per year." He adapted his own famous 2004 Democratic keynote speech—wherein he declared that only pundits, spin masters, and negative ad peddlers would "slice and dice our country into Red States and Blue States"—to echo Occupy language:

> I believe that this country succeeds when everyone gets a fair shot, when everyone does their fair share, when everyone plays by the same rules. These aren't Democratic values or Republican values. These aren't 1 percent values or 99 percent values. They're American values. And we have to reclaim them.

President Obama and his advisors monitor the Occupy movement and make tactical decisions about how to relate to it. But it's almost certainly too late for him to rouse the Occupy hard core. For good reasons and bad, he has foregone their trust, and winning it back would be exceedingly hard even if he tried, for he is in no position to deliver much concrete reform against Republican obstruction, and rhetoric won't fill the bill. Many, probably most of the inner Occupy movement—including some who worked for Obama in 2008 but stayed home in 2010—are tired of lesser-evil arguments, even of reminders about Republican obstruction, the danger of giving a Republican president more Supreme Court nominations, and so forth. They have not forgiven Obama for appointing Tim Geithner and Larry Summers. They have moved on. Some Occupy activists do understand that the prospects for organizing, and winning real improvements for people's lives, are better if Obama retains the White House, but some don't. If they turn out for him at all, it will be defensively, with expectations so low they graze the dirt. Some Occupy supporters will grimace and vote for him, but some, especially in New York and other strongholds of the left, may not turn out at all.

During the campaign year, the greatest danger facing the movement and its chances of overcoming the plutocracy is the prospect of blood in the streets. Three prime opportunities for protest would present themselves: the first in mid-May, during the combined G8 and NATO summit meetings in Chicago; the second, at the end of August, during the Republican convention in Tampa, Florida; the third, at the beginning of September, during the Democratic convention in Charlotte, North Carolina. The Chicago spectacle would not, strictly speaking, be a campaign event, but it would be a media magnet and any collision on the streets—anything more enthralling to the cameras than the pomp and circumstance of foreign ministers congratulating each other, in fact—necessarily would become an event in the national campaign.

The thought of confrontation in the streets of Chicago necessarily sent the mind racing back to August 1968, when Mayor Richard J. Daley militarized the city in the run-up to the Democratic Convention, and in unstated (and in some cases, thanks to *agents provocateurs*, de facto) alliance with the most reckless or insurrectionary protesters precipitated a confrontation that, regardless of the fact that the great majority of physical force came from the police side, helped consolidate public opinion against the antiwar movement, and usher in a hard-headed (and hard-hatted) president as a law-and-order man, even as the Vietnam war itself was declining in popularity. In May 2012, Chicago Mayor Rahm Emanuel decided to understudy the twentieth-century boss, proposing huge new expenditures for surveillance, large fines for minor crimes, the purchase of million-dollar bonds to insure against property damage, time limits for demonstrations, advance registration of any objects to be carried, amplification limits, and a host of other restrictions. As the longtime Chicago political observer Don Rose pointed out, few if any of those issues would be

in the control or capability of the organizers of the G8/NATO protests. These events historically draw crowds from all over the world. No person or organization is in charge of all the protestors, nor can anyone predict how many people will show up, how many marshals will be required, who will or will not have a bullhorn or trumpet, how long it will take for everyone who has something to say to speak their piece nor how long it will take a crowd to march from the staging area to the site of the meetings.

Emanuel was "building in failure to comply with the new laws and putting all the protestors at risk of arrest and huge fines." Chicago aldermen and newspaper editorials protested some of Emanuel's stringent proposals, and the city government pretended to take the protest seriously and began issuing permits.

Then, in March, the White House took command. Well aware that handfuls of disciplined militants could smash things and grab the spotlight if they wanted to, and that savage police could do likewise, the Obama administration adroitly announced that it was moving the G8 meeting to the "more casual" high-security woods of Camp David, Maryland—thus defusing a political time bomb, though what would come of Chicago's NATO meeting remained to be seen.

The Obama campaign was acutely mindful that in the wings of the larger movement awaited the sorts of heavy-duty anarchists who in Europe and Oakland, at least, seek opportunities for high-profile window-smashing, flag-burning or (in Italy) car-torching. These so-called black blocs seized screen time in Seattle in 1999, in Genoa in 2001. It is far from clear exactly who the vandals are, even whether they are regulars or occasional participants, for they do not hold press conferences or send out press releases or submit to interviews. But in any case, they are the camp followers of larger, pacific assemblies. (A less kind term would be *parasites* in the strict sense, for they cannot thrive except on the bodies of larger hosts.) There are the hundred or so masked window-smashers who detached themselves from a peaceful general strike in downtown Oakland on November 2, 2011. Among them are a few insurrectionist anarchists in New York (one number bruited about is twenty), who find inspiration in Italy and Greece. Fortunately, few OWS radicals are enamored of them. To a radical activist like Meaghan Linick of Organization for a Free Society, "They write poetic manifestos that don't make any fucking sense." Or they import them—for example, a spirited French tract of 2005 called "The Coming Insurrection," whose arrogantly marginal flavor is conveyed by its opening:

> From whatever angle you approach it, the present offers no way out. This is not the least of its virtues. . . . Everyone agrees that things can only get worse. "The future has no future" is the wisdom of an age that, for all its appearance of perfect normalcy, has reached the level of consciousness of the first punks.

Though on paper the marginal insurrectionists argue fine points—does property damage actually constitute violence? is violent self-defense permissible?—they are not averse to courting instigations, especially when the police abuse their authority, deploy high-powered weaponry, and shed blood, tear gas, and pepper spray. The insurrectionists and the most aggressive police are in a silent alliance. When orange plastic nets are used to seal off peaceable assemblies, confining them to so-called protest zones far from the objects of protests, even nonviolent activists will be enraged at being corralled (*kettling*, it has been called) and tempted to abandon their discipline. It was the kettling of two peaceful protesters on a sidewalk in Greenwich Village, and their screams in reaction, that led an overwrought police officer to pepper-spray them in the face on September 24—and, by the way, provide bad press for the police and a big publicity boost for the Occupy movement.

As for the party conventions, the city governments of Tampa and Charlotte have also raised their heavy hands, helped by Homeland Security purchases of high-tech equipment. Orange netting, massive armament, and surveillance are orders of the day. Tampa's police department boasts a new armored SWAT truck along with two aging tanks. Charlotte banned overnight camping and, according to North Carolina ACLU legal director Katy Parker, under a proposed new ordinance, "Officials can wait 20 days before deciding whether to grant a protest permit, decide how many police are needed to oversee the assembly and then charge demonstrators for police and fire costs." One has to admit that there is a certain ingenuity in the idea of user fees for those who would avail themselves of the First Amendment. Talk about the marketization of all value. Talk about abject surrender of the premise of the First Amendment: that freedom of assembly is an irreducible value in the republic of the United States of America. Where are the originalists when you need them?

During the winter of 2012, most Direct Action activists were too busy deciding on actions to undertake during the following six months to think very clearly as far ahead as the party conventions, though some clearly wanted direct actions there. There was talk of counter-conventions to discuss issues, and even of a permanent occupation site somewhere in the East, to serve as a base for convention-related actions and even for continuing encampment. About what would happen then, there was a certain let-the-chips-fall-where-they-may attitude. "We don't know what [city officials] are going to do," Yotam Marom told me, "They'll do what they're going to do." In January, Ed Needham of the OWS Public Relations Working Group anticipated that:

> The conventions will be tinderboxes. Given the lack of long-term strategic planning, I'd be very surprised if there were no confrontations with the police at the conventions. Of course there'll also be nonviolent actions focused on issues. Both will happen simultaneously. Those will be times of great struggle for the movement.

Would the main body of the movement be ready to cope with outbursts of violent protest? I asked him. "I don't know. I don't know if the Democrats will be ready either. We will have to face these forks in the road. We will have to constantly ask ourselves who we are."

On September 23, New York's GA reached consensus on what it called Principles of Solidarity, declaring: "We proudly remain in Liberty Square constituting ourselves as autonomous political beings engaged in non-violent civil disobedience and building solidarity based on mutual respect, acceptance, and love." These words are still in force as I write. "Nonviolent civil disobedience" would seem to settle the matter—on paper. But the police and those who command them will have a good deal to say about what develops on the streets, and on the screens that convey images of the streets while the proverbial whole world watches. Cage peaceful demonstrators who, by putting their bodies on the line, take seriously the right of the people peaceably to assemble, and some of them will be sorely tempted to get out of line. Then, on the battlefields of public imagery where hearts and minds are won and lost, it scarcely matters who casts the first stones, smashes the first windows, or physically attacks the first police or wealthy executives. It is not enough for the movement to be *mainly* nonviolent. The movement as a whole needs to anticipate damage, try to contain it, and prepare damage control.

For this reason, I agree with those within the Occupy movement who want to renounce violence more explicitly and formally, as well as step up nonviolent training. "Diversity of tactics" is a pleasingly pluralist ideal, tantalizingly broad enough to embrace anything from the right to fight back against police violence to the-sky's-the-limit. In the real world where the movement's interior process matters rather less than the headlines, it is solipsistic. Since violence and arrests, whichever come first—and flag-burning most of all—always commandeer the media spotlight, even those who oppose violence need to recognize that the failure to renounce it explicitly passes a publicity cudgel to those who would succumb to Joseph

Conrad's exhortation from *Lord Jim:* "In the destructive element immerse!"

On November 8, 2011, three committed activists of the Alliance of Community Trainers posted an open letter advocating that Occupy adopt a "framework . . . of strategic nonviolent direct action." What they meant was that before any given action:

> Occupy groups would make clear agreements about which tactics to use. . . . This frame is strategic—it makes no moral judgments about whether or not violence is ever appropriate, it does not demand we commit ourselves to a lifetime of Gandhian pacifism, but it says, "This is how we agree to act together at this time."

Their arguments were strong in their own right, though they failed to address the media-violence nexus explicitly. In the spirit of a movement that values inclusiveness, they pointed out that violence, or even the refusal to renounce violence, is in effect exclusionary:

> Lack of agreements privileges the young over the old, the loud voices over the soft, the fast over the slow, the able-bodied over those with disabilities, the citizen over the immigrant, white folks over people of color, those who can do damage and flee the scene over those who are left to face the consequences.

They argued, moreover, that "lack of agreements and lack of accountability leaves us wide open to provocateurs and agents," and that this wide-openness worked against the movement value of transparency and accountability. They acknowledged that:

> Not everyone who wears a mask or breaks a window is a provocateur. Many people clearly believe that property damage is a strong way to challenge the system. And masks have an honorable history. . . . But a mask and a lack of clear expectations create a perfect opening for those who do not have the best interests of the movement at heart, for agents and provocateurs who can never be held to account. As well, the fear of provocateurs itself sows suspicion and undercuts our ability to openly organize and grow.

One comment on their post, a month after it went up, was especially encouraging, from someone calling him- or herself Cameron:

> I have been deeply involved in the Occupy Movement in Oregon and Washington, mostly in Seattle, since the beginning of October. In my past, I have thrown bricks through windows and gotten in physical altercations with police officers. I am not afraid to commit such acts. However, for reasons that seem obvious to me, tactics that employ such means could not possibly be effective in achieving the lofty goals the Occupy Movement aims for.

As yet, New York's General Assembly has not addressed the proposal. Tensions over whether the movement should do so in some fashion were running high.

It couldn't be plainer that Occupy turns its collective back on political endorsements. The culture of the Occupy movement clashes with the culture of elections. For the movement, a refusal to endorse transcends any question of strategy; it's a matter of identity. The strategic rationale is articulated by Bill Dobbs, who brings to Occupy's press team decades of experience with ACT UP and the movement against the Iraq war:

> If things are going to shift, my take is that sustained organizing around issues and being a squeaky wheel has a shot. Organizing around candidates, parties, elections will not do it. That means not ignoring electoral politics or the people in power but keeping them jumping. Staying at an arm's length rather than discovering that a kiss on the cheek from a politician quickly turns into a suffocating bear hug.

But this is not to say that the movement's hands are tied during the 2012 campaign. During the primaries, activists set out to "occupy the agenda" (the journalist Sarah Seltzer's phrase) in Iowa, New Hampshire, and other states, mic-checking the candidates and injecting economic inequality issues into town meetings. One candidate from the Occupy Philadelphia announced that he would run in the Democratic primary against a conventional incumbent, though he later withdrew when his qualifying signatures were challenged. Meanwhile, regardless of what the more electoral-minded Occupiers may do on their own—register voters, knock on doors, make phone calls, the whole universe of tasks that go into winning elections—some activists argue that Occupy might leave its most lasting imprint on American politics by stepping outside the candidate runs and focusing on deep constitutional changes that in the longer run might just barely be attainable. The idea circulates for a campaign for a constitutional amendment that would provide for complete public financing of elections and ban private donations over, say, twenty-five hundred dollars.

Advocates of this tactic are not altogether naïve. They are aware of how arduous the amendment process is and how few squeak through the gates of ratification. Since the initial ten that make up the Bill of Rights, new amendments have been ratified, on average, only once every twelve years. Since several GAs have already endorsed resolutions opposing corporate personhood—the prevailing fiction that corporations are entitled to the protections guaranteed individuals—it's not unreasonable to think that they might pour energy into an amendment campaign that extrapolates from that very principle. By January, David DeGraw, who could lay claim to having devised the "99 percent movement" slogan in the first place, and since then busied himself building the Interoccupy network (interoccupy.org) that linked six hundred Occupy locations across America, thought that groups outside New York were warming to the idea of making demands and otherwise working within the political system. The notion of a constitutional amendment to get money out of politics was, he thought, "the most 'consensed' issue" in the movement. Some advocates of an amendment campaign, interestingly enough, had money to spend on it. He reckoned that the only Occupy people who disagreed were the roughly one-tenth who didn't want anything at all to do with the political system. The strongest contingent of those was in New York: "People are more headstrong in New York," he told me.

The central pillar of plutocracy is the domination of politics by money. On this there is broad popular agreement. Whether a fog of agreement can be condensed into a torrent of political energy is—given the cumbersomeness of the American Constitution—quite another question. There will be no way to know without years of work. Years.

15. Can the Outer Movement Get Organized?

If Occupy's inner movement is otherwise occupied, unlikely to coordinate on a political strategy in the foreseeable future, what of the outer movement—the trade unions, liberal lobbies, and professional groups, the Congressional progressives, all the membership and Washington-centered organizations that specialize in federating, horse-trading, infighting, and at times getting practical things done? Unions lent Occupy a good deal of material support—feeding campers, donating portable toilets, paying organizer salaries—and rallied tens of thousands for large marches (October 5, October 15, and November 17, largest among them) that impressed much of the established as well as the unestablished media with the movement's reach. (But true to form, media could not be relied on for accurate depictions of the movement's activities, whether candlelight vigils or eruptions of performance art. Disparagement did not end when labor and mainstream liberals widened the movement. When CBS correspondent Jim Axelrod claimed on a live feed to the national news that there were only "a few hundred" demonstrators in Foley Square around 6 PM on November 17, and I could see with my own eyes that there were tens of thousands, I was reminded that journalistic malpractice never vanishes altogether. Even after Occupy soared nationally, not everyone was impressed, and a narrative of incipient decline was rarely far out of reach.)

If the outer movement sees Occupy as too far outside the system, incapable of or uninterested in organizing coordinated pressure for economic reforms like government job creation, a Wall Street transaction tax, and tax surcharges on the ultrawealthy, isn't it up to the outer movement itself, on its own, to promote a reasonably attainable program and mobilize to attain it? Why shouldn't they do their own thing, consolidate their alliance, be grateful they have Occupy as an inspiration, make sure to refrain from speaking in its name, and, well, move on? Will the outer movement groups that meet to coordinate agree on such a common demand as what one of them calls a "21st-century Glass-Steagall Act," a bill to break up the nexus of investment and commercial banking that permitted the banks to turn into casinos gambling with securitized mortgage bundles? As members of Congress introduce such laws, can they mobilize numbers to support them? If it's up to the radicals, among others, to make the Occupy movement endure so that the respectables are forced to take the need for substantive reform more seriously, isn't it up to the respectables to prove the radicals wrong about whether the American political system can reverse decades of growing inequality?

As it is, such a sympathetic unionist as Marty Frates, secretary-treasurer of Oakland's large Teamster Local 70, worries that Occupy's fuzzy and inconsistent messages, and its attachment to radical anarchist ideals, may preclude its playing the part it needs to play in the elections. That is the nature of the Occupy beast, at least in the militant-laden Bay Area. Oakland Teamsters at least engaged the Occupy activists—attended meetings, made arguments. In Chicago and elsewhere, activists returned the favor by supporting militant workers. But overall, even as national labor leaders issued ringing endorsements of Occupy, their organizations were otherwise engaged. There was, for one thing, a certain reluctance to be perceived as interlopers and come-latelys. Call the reason conscience, modesty, weakness, or lack of vision, the outcome was the same—the failure of respectable organizations to coalesce around a set of economic reform demands and mobilize accordingly. Sitting politicians are, of course, less bashful—taking credit is the name of their game.

Historically, coalitions of outer-movement and inner-movement groups have accomplished what individual groups could not. In the fifties and sixties, the Leadership Council on Civil Rights brought together all the civil rights groups, from the most militant to the stodgiest, along with religious and union leaders, and organized liberals, to coordinate work, to lobby, debate and coordinate strategy, and make deals with the Kennedy and Johnson administrations. The Leadership Council included experienced men (though few women), many of them visionaries with strong followings outside the Beltway: Martin Luther King, Jr. of the Southern Christian Leadership Conference; James Farmer of the Congress of Racial Equality; John Lewis of the Student Nonviolent Coordinating Committee; Roy Wilkins of the NAACP; Whitney Young of the Urban League; Walter Reuther of the United Automobile Workers. There is today no equivalent, muscular body, no Leadership Council on Economic Rights. Instead, there is a scatter of websites and programs. The AFL-CIO's community affiliate, Working America, offers a "9 Demands for the 99%" website. (It lists eight economy-based demands and, in a participatory gesture, invites readers to suggest a ninth.) The Congressional Progressive Caucus offers "Restore the American Dream for the 99% Act," which carries an overlapping agenda. Many lists of relevant demands are hiding in plain sight, but no master coalition has emerged to promote them.

Meanwhile, the unions are weak and getting weaker. It's no secret that they have been steadily shedding members for years—from 20.1 percent of the work force in 1983 to 11.9 percent in 2010. Workers aged fifty-five to sixty-four are the most unionized (15.7 percent) and those 16 to 24 the least (4.3 percent). More than one in three public sector workers are unionized (36.2 percent) as against less than one in fourteen in the private sector (6.9 percent). The unions are embroiled in self-maintenance, ensnarled in jurisdictional disputes and administrative battles. Visionaries are scarce. There is no Walter Reuther.

Occupiers in union-heavy New York noticed that, in October, the Transport Workers Union protested having their buses commandeered by the police to drive Occupy arrestees to jail, and that on October 5, October 15, and November 17, when unions turned out their members, the Occupy marches swelled by orders of magnitude, from a thousand or so to tens of thousands. On the other hand, when labor marched on its own, it looked and sounded lackluster. On the left coast, some younger union activists threw themselves into Occupy Oakland, went to meetings, mic-checked, and twinkled. Even when they kept their distance, unionists thought Occupy was cool. On November 2, after police ran amok against the local Occupation, unions made donations. However, when Occupy camps decided to blockade west coast ports on December 12, to "block the 1 percent" and cut into their profits, thus "taking on the global issue of exploitation by capitalism," as one Occupy Oakland organizer put it, even if the blockade cost the City of Oakland revenue that supports the city's social programs, the unions objected, and not only because they were legally bound to do so if their members weren't to forfeit a day's wages. The secretary-treasurer of the Alameda Building and Construction Trades Council, Andreas Cluver, said:

We're extremely supportive of the message of Occupy Oakland, and we did come out to support the November 2 general strike, but we're not behind this one. When working people aren't involved in the decision on whether to shut down their jobs at the port, that's problematic. And we weren't consulted. Losing a day of wages is hard.

Occupy had stepped squarely into labor's major cleavage—between the organized and the unorganized. Some longshoremen dissidents supported the December 12 port shutdown, but the main worker support came from truck drivers who drive the unloaded goods out of the ports. A group of them issued an open letter saying:

> While we cannot officially speak for every worker who shares our occupation, we can use this opportunity to reveal what it's like to walk a day in our shoes for the 110,000 of us in America whose job it is to be a port truck driver. . . . We feel humiliated when we receive paychecks that suggest we work part time at a fast-food counter. Especially when we work an average of 60 or more hours a week, away from our families. . . . The companies we work for call us independent contractors, as if we were our own bosses, but they boss us around. We receive Third World wages and drive sweatshops on wheels. We cannot negotiate our rates. (Usually we are not allowed to even see them.) We are paid by the load,

not by the hour. So when we sit in those long lines at the terminals, or if we are stuck in traffic, we become volunteers who basically donate our time to the trucking and shipping companies. That's the nice way to put it.

One signatory, Leonardo Mejia, of Long Beach, told me more. His is a tale about the 1 percent, in fact, for his employer, SSA Marine, the largest US-owned container terminal operator, is 51 percent owned by Goldman Sachs. Mejia started driving trucks at the ports in 2001. He said:

When all the port drivers used to own our own trucks, the situation wasn't so bad. It changed in 2008. At that point, most of the port drivers were making twenty-eight to forty thousand dollars a year before taxes. The ports created a plan to avoid pollution. They forced us to sell our trucks and buy new trucks, and the state paid the ports to buy new trucks, thinking that the truck drivers would be recognized as employees. But only fifteen out of 850 companies in Los Angeles have recognized the drivers. If I want to keep working at the port, I have to drive a new truck. Now most of the companies force drivers to lease the truck if they can't buy it. You have to pay all the insurance, the diesel, the maintenance—you have to pay everything. So in the past year, I made eighteen thousand dollars. I have to pay three thousand dollars in taxes to the IRS. I work between nine and eleven hours per day, six days a week or as many as the dispatcher says. I have no control. If I want to go to work for somebody else, I can't do that. I must work only for that company. They say that we're independent contractors but in reality we're employees—

employees without benefits, that is.

Those who defend the millions of dollars paid America's wealthiest executives on the ground that they earn their "compensation" because they work hard might like to tell the world in what way they work harder than Leonardo Mejia. The corporate practice of labeling controlled employees "independent contractors" flies in the face of IRS regulations that technically distinguish between the two. But when the government is managed for the benefit of capital, it comes as no surprise that this deceptive practice is widespread. It also comes as no surprise that the business press pays little attention.

Who benefits from such deceptions, then? It must be acknowledged that the 1 percent have allies, collaborators, and beneficiaries. An executive of SSA Marine told Bloomberg News that Goldman Sachs Infrastructure Partners "is primarily made up of pension plans of workers in the United States and Australia, and those groups hire money managers to manage their pension funds." In his eyes, "We are a union operation; we support union workers, family-wage projects and make investments to increase those job opportunities." SSA Marine maintains, in effect, that it impoverishes truckers in order to benefit the part of the working class that enjoys pensions.

Mejia is an optimist. The fight for union recognition has been underway for decades. He thinks it is destiny that Occupy came along. "I don't know if you believe in destiny," he says, "but I do. The process is slow—is going to take a few more years. But we're going to win. Occupy is fighting for a just cause. If we join forces, we can win because we have the same enemy. Definitely we have to do more actions together. I don't know when, I don't know how, but we need them and they need us."

16. Is There a Global Revolution?

Vast changes do not neatly follow the calendar, but it is already possible to say that the year 2011 was, as Anthony Barnett writes, "original."

Not completely so, of course. As in 1848, 1968, and 1989, the insurgencies were many and they absorbed multitudes. As in all three, the protagonists were chiefly young. As in all three, the holders of power felt various degrees of panic. As in 1848 and 1968, uprisings took place on more than one continent. As in 1968 and 1989, the insurgents were largely nonviolent, until the uprising in Libya. As in 1968, the targets were multiple, the identities of the movements alternately seductive and repellent in the eyes of outsiders, and often confusing.

The grandest originality was that in contrast to 1848's uprisings across Europe and Latin America in behalf of nationalist and republican values against absolutist government and economic impoverishment, 2011 was chiefly nonviolent. The second, of course, lies in the electronic means of communication: the smartphones, videos, social network and other internet linkages that sent the horrific images of the self-immolated Mohammad Bouazizi and his funeral procession flying throughout Tunisia; then the photograph of the mutilated face of the twenty-nine-year-old Egyptian businessman from Alexandria, taken by camera phone in the morgue by his brother, around which formed the momentous Facebook page posted by the Google executive Wael Ghonim, "We Are All Khaled Said," circulating throughout Egyptian cyberspace, along with the call to gather in Tahrir Square on January 25, so that cyberspace touched down on earth, and in the flesh, face to face, groups formed, found their affinities, intermingled, sized up their situations. Graphic images have become more graphic and they move faster; they horrify instantaneously. The cascades of images, horizontal contacts, and related events have sped up enormously. But this most visible of differences from past revolts can be exaggerated. Before there were online videos, there were gossip networks, secret societies, broadsides, posters, leaflets.

There were mimeograph machines, printing presses, the mails and telephones. Forty-six and a half years before the Chase Manhattan Plaza of 2011 became Occupy's first-choice target for the hundreds who rallied at Bowling Green, on March 19, 1965, Students for a Democratic Society organized a sit-in at the Chase Manhattan Bank protesting loans they had made to bail out the apartheid regime in South Africa after it had massacred sixty-nine demonstrators and injured 180 (many of these shot in the back) in the township of Sharpeville five years before. How did SDS communicate its projects then? Lumberingly, and yet with some effect, allowing for different circumstances. In 1965, SDS's national office mailed a monthly mimeographed bulletin to its few thousand members, along with a bi-weekly "Work List" bulletin, also mimeographed, which went out to some one hundred of the most committed activists. Long-distance calls were expensive, and so kept to a minimum. SDS barely bothered with press releases. There were no underground papers, no cable news, no blogs or smartphones. Yet information spread and things got done. Around the country, a dozen or more demonstrations over corporate collusion in apartheid got coordinated. On March 19 itself, the later-to-be writer Mike Davis and I, skulking around Chase Manhattan in our versions of bankers' garb, were able to use an ancient instrument called a pay phone to inform the prospective sit-inners, awaiting our last-minute signal in a nearby office, exactly which bank entrance to block (to maximize our chance of positioning ourselves on the sidewalk in front of the revolving doors before the police could obstruct us). Forty-one were arrested while several hundred picketed and shouted up on the plaza.

The press was not awfully interested in this eruption on Wall Street, unusual though it was. Still, it did draw fifteen column-inches on page 11 of the next day's *New York Times,* in a bare-bones account that devoted four paragraphs to a bank official's denial of malfeasance but lacked any reference to SDS's research into the role of American banks in shoring up the apartheid regime, or to the bank's attempt—a failure, as it turned out—to get a judge to enjoin any reference to "Chase Manhattan, Partner in Apartheid" in SDS leaflets and research papers and buttons passed out during the previous weeks. Still, though SDS moved on to antiwar work, the issue of corporate complicity in apartheid was taken up by a host of church—and Africa-related groups—a modest cause that developed over the next decades using modest communications.

Two very different demonstrations, moments in two very different movements, and not only, not even chiefly, because the means of communication had evolved from mimeographed mailings to Facebook groups, Twitter tweets, camera phones and free conference calls, but because the whole surrounding society and the protest issues had changed, changed utterly.

The sluggishness of the past is an illusion. So is the isolation of history-makers from one another. So is the inevitability of liberation, at least in any conceivable short run. In 1848, the revolts were crushed, and resulting reforms scanty. In 1989, Mikhail Gorbachev's reforms ignited Central and East European revolts that the Communist apparatus could not contain, though the Chinese, willing to open fire, could and did.

The year 2011, as Anthony Barnett observes, saw an extraordinary discrediting of political-economic elites on both sides of the Atlantic: "The two fundamental ideological pillars of the North Atlantic order, that it keeps the peace and that it delivers wealth for all, are clearly broken." (Or actually, in the words of one OWS sign observed in Albany, New York: "The system isn't broken. It's fixed.") It wasn't that bread was no longer baked and streets no longer cleaned (some of the time, anyway), but that damage from the crash of 2008 continued to rain down from the commanding heights. Vast hurts were not undone and structural instabilities were not remedied. Once again, Anglo-American financial leadership embraced market fundamentalism, precisely the ideological regime that had produced the crash in the first place. "When the bubble burst," Barnett writes, "this *exposed* the political system as a 'post-democracy' that answers to corporate power." The moral authority of both political and economic elites was at low ebb.

So was the efficacy of state violence, as American troops left Iraq, mission unaccomplished. Both Osama bin Laden and the war on terror expired in 2011. In the Arab world, Barnett observes, "The abject failure of terrorism and its evident pointlessness and monstrosity . . . opened the way to peaceful, popular uprisings," It is poetic—but not only poetic—justice that Zuccotti Park, the center of an extraordinary social invention, is a stone's throw from the site of the momentous destruction of the World Trade Center. And it was not only bin Laden's achievement as "hooligan of the absolute" (Tom Nairn's phrase), but George W. Bush's "mad logic" of "hugely inflating the significance of its enemy" that was discredited, along with the tyrannies that were supposed to hold the fort in North Africa. "The 'War on Terror'—far from protecting the world or securing US hegemony—proved to be chasing a global chimera," Barnett adds. "There is terrorism, but it is a very nasty criminal danger not a strategic threat." The elites' political capital was much depleted. And in the United States, at any rate, rallies around the flag were no longer capable of smothering opposition.

It would be shallow to see uniformity in the global crisis of legitimacy, or the popular reactions to it. The anti-austerity protests in Greece, Spain, Ireland, and elsewhere have their respective characters and limits. Islamists, who are at least short-term victors in North Africa, do not have equivalents elsewhere. However, it would be equally shallow to expect uniformity, let alone demand it in the old mythic form of "the revolution," in the singular. Theorists have been proclaiming this the age of late capitalism for decades, but if it is indeed late, it keeps getting later. Capital continues to find new ways to grow, creative destruction and all. We are not witnessing the global *Götterdämmerung* of the political elites or their institutions. They persist if for no other reason than that no new world is yet emergent within the shell of the old. But it is fair to see 2011 with Anthony Barnett as both politically and chronologically "the end of the decade that began with 9/11: from the year of the towers to the year of the squares."

17. "This Is the Beginning of the Beginning"

Is it auspicious for the movement's prospects that, as of early 2012, supermajorities still weigh in for substantial reform? It would seem so. A recent Pew poll tells the tale:

> Roughly three-quarters of the public (77%) say that they think there is too much power in the hands of a few rich people and large corporations in the United States. About nine-in-ten (91%) Democrats and eight-in-ten (80%) of independents hold this view; a much narrower majority (53%) of Republicans do as well. For historical perspective, six-in-ten (60%) Americans expressed this view in a 1941 Gallup poll

[i.e., twelve years after the onset of the Great Depression—TG]. Reflecting a parallel sentiment, 61% of Americans now say that the economic system in this country unfairly favors the wealthy.

However, Nona Willis Aronowitz is right to point out that, according to the same poll, just about as many Americans think that the rich got rich "mainly because of their own hard work, ambition or education" (43%) as "because they know the right people or were born into wealthy families" (46%). These results have not varied significantly since 2008. Half of America thinks that it lives in a class society, that is to say, in a society that's rigged to favor a few. The other half thinks it lives in YOYO America—You're on Your Own—and that it is the best of all possible Americas to live in, or at least the only imaginable one.

Poll numbers do not necessarily catch the deeper currents that move people beneath the surface, or that may just be stirring, even below consciousness, at levels of feeling and thought way beneath opinion, attitude, or even what is conventionally called *belief*, and so elude pollsters. What moves people in history is often the more subtle matter of which of several conflicting values will come to the fore. It is not just a matter of what they want in the abstract, but of whom to trust, to affiliate with, to hang out with, to be moved by—and what, in specific circumstances, to do. Statistical supermajorities do not, by themselves, move history, and even real ones have to be knit together into working alliances of disparate populations, and moved in turn. Favorable conditions and auspicious prospects do not necessarily make for wise decisions by a movement that is making itself up as it goes along, while circumstances change and adversaries adapt. The thrill of talking about fundamental reform or radical change—of sharing a feeling that the world has already changed, that doors have been flung open, because this kind of talk is now possible—is a necessary, but far from sufficient, condition for accomplishing fundamental reform or making radical change.

Just so, the movements of the sixties produced major cultural and political changes, advanced the democratic project in countless ways, but their radical potential was blunted and at times reversed by sloppy thinking, reckless posturing, and mindless violence that played into the hands of our adversaries. In the grip of furies, some of the most passionate activists did not seize the moment—to consolidate and to think—but were seized *by* the moment, by its incandescence and wildness, in the belief that by sheer force of will, with the aid of magical incantations that they mistook for revolutionary guides, they could bull their way through weakness and uncertainty. So, in the caldron of the late sixties, it came to pass that just as the revolt against illegitimate authorities was spreading to vast new populations of women, gays, and others, and just as questions about American values were churning everywhere, the possibility of a more extended left-of-center majority was wrecked by the desperate overconfidence and extravagant fantasies of the Black Panther Party, their offshoots and rivals, the

Weather Underground and other microfactions, and by hard-drugging utopians who failed, or refused, to understand that there were vastly more Americans who occupied ideological territory to their right than to their left, and imagined that they could go it alone into a revolutionary promised land.

After the October 5, 2011, march from Zuccotti Park—Liberty Square—to New York's courthouse center at Foley Square, the first time an Occupy turnout climbed over ten thousand, a neighbor said to me: "Beginnings are always wonderful."

Her smile carried some ruefulness. Veterans of social movements know well the cycles of hope, energy, hope against hope, success, failure, expansion, implosion, desolation, recalibration. . . . Disappointments follow appointments. It's in the nature of ragged, zigzag history—the only kind there is—that movements flow, ebb, churn, even change streambeds. There are no guarantees, not even in the short run. Emergencies are as common as urgencies. In my days as a college activist—in a little band possessed of the chutzpah to try to get nuclear weapons under control—we frequently worried that the current year's freshmen (as we called them) weren't showing up the way they had the previous year, that the last demonstration wasn't as impressive as the next-to-last, that our ideas were stale, that internal tensions would wreck us; that we had no good follow-up to the last action. (Half a century later, we still strive to get nuclear weapons under control.) In the civil rights movement, the antiwar movement, the women's movement, there was always a fear that successes meant symbolic absorption, that the next action could easily go wrong, that the movement was sinking, stagnant, stalled, stuck, riddled by destructive dissension. (Sometimes it was.) Wall Street likes to say: The market climbs a wall of worry. Movements climb walls of worry. And sometimes fall from them.

The Occupy sign that reads, "This is the beginning of the beginning" is of course hopeful, and hope is audacious. It does not require any guarantee of a happy ending. However, it is also true that those who dare hope do not absolve themselves of the ordeal of having to live, along with many others, with the consequences of their actions. They are not the only force at play, but the movement is the force they themselves can steer.

So, Occupy flows, meanders, and tenses. "It," in fact, is not "it." It is, at least at times, a verb in search of nouns. The movement moves. It knows that it needs to be not only vigorous and multiple but smart. At its best, it learns. One of its achievements, in the early months, was to become, at least sometimes, its own school. It learned from itself. It processed its own experience. In the sometimes viscous flow of its awkward assemblies and working groups, it could entertain shallow ideas and shrug them off, just as it could sometimes succumb to them. The movement could learn, not always brilliantly, to be an encampment, and it could cope with being evicted, though there were some for whom the means (occupation) turned into the end (challenging plutocratic rule). For example, there were those in the movement who thought that the most important thing was finding a space to occupy, so that when, after the November 15 eviction, Wall Street's Trinity Church—which is wealthy, but is also a church—refused to turn over a nearby parcel of unused, fenced-in land to be occupied, they mobilized to get over the fence on December 17 but were able to summon no more than three hundred to their rally.

Conceivably, Occupy could evolve into an enduring force—an awakening that, like its predecessors in the eighteenth and early nineteenth centuries, and in the sixties, irreversibly changes the values that Americans live by. Conceivably the movement could recruit new adherents while keeping the veterans excited. Or conceivably, it could devolve to the point where it puts out calls for actions and people stop showing up. Conceivably, it could build enduring bridges to membership organizations, and engage the political system in such a way as to keep its principles clear and at the same time win results. Conceivably, it could become obsessed with confronting the police while lots of erstwhile supporters sigh and pine for a transformation that might have been. Seized by their paramilitary mission in a Homeland Security state, the police, after all, can be expected to offer the movement many more occasions for outrage.

The Occupy movement is an obscure object of desire and fear. Its whirlwind emergence out of apparently thin air has upended so many expectations, revamped so many people's sense of possibility, that people who know little about it, or see it as the heart of their desire, have ideas about where it should go. From the start, well-wishers joined not-so-well wishers in saying that Occupy ought to do X, Y, or Z. Everyone is a critic—insiders, outsiders, all. At a symposium on the Occupy movement at Columbia University, political scientist and longtime movement strategist, Frances Fox Piven, made a decisive point. It was misguided, she said, to obsess about what Occupy should do. The questions are what will *you* do, what will *we* do.

A serious social movement is one that makes the political personal in a thousand ways, even as a core of the most passionate decide whether and where and how to take to the streets. Beneath the outcroppings of visible public events lies a movement's secret life, where people think and talk around their kitchen tables. Perhaps they have been swept up in a high school demonstration, a candlelight vigil, an anti-eviction blockade, an Occupy-inspired election campaign. Some identify with the movement; some don't. Some like its spirit; some are suspicious. Some identify with one tendency; some with another. In the sixties, millions of people asked questions like: What do I do about racist terror? Do I obey unjust laws? Do I go to a rally? Do I consent to be drafted? One way or the other, what do I say and do afterward? Am I willing to get arrested? To work for a referendum? To go into politics? Should I keep my job in the military-industrial complex? Women asked themselves, should I leave my husband? Gays asked, should I come out of the closet? Doctors asked themselves where they should practice; graduate students, what they should study; soldiers and sailors, what they should do in an unjust war. Millions of people thought themselves over. They asked themselves—to paraphrase an American president—not what the movement should do for them but what they should do for the movement. One test of the Occupy movement is whether it will continue to inspire questions like these: If I have a degree in physics or mathematics or, for that matter, economics, do I want to go to work for Goldman Sachs? Can I put my online knowhow to work organizing antieviction squads? If I am an architect, do I want to design buildings for the 1 percent?

The sign, "This is the beginning of the beginning," is exactly right. The question is what do *we* do.

AFTERWORD

In 1989, Shen Tong was a leader of the Chinese democracy movement in Tiananmen Square. After the massacre, he escaped to America, studied first biology and then political and social theory, became a software entrepreneur, and founded a multinational company headquartered in lower Manhattan. Zuccotti Park called to him, and he threw himself into the Occupy movement. He became a prime mover in a "messaging cluster" that strove to consolidate the movement's attention on specific objectives whose hallmark would be economic fairness and justice, helping form a working group called 99solidarity.net toward that end. He raised funds for Occupy. He is an extremely well-organized man in his early forties—which he needs to be, what with an ever-full inbox and young children. (When babysitting arrangements fail, he has been known to bring his baby daughter to a working group meeting.) Shen Tong likes to say that a social movement can be crushed by two kinds of crisis: one kind is a massacre; the other is success.

The success he refers to is the transformation, within a few weeks, of America's political discourse. The movement's friends and its enemies could agree that there was a new center of gravity in what we are pleased to call "the national debate." Inequality of wealth was now widely recognized—and seen as a problem, not a natural condition. "The 1 percent" and "the 99 percent" were commonplaces. Politicians felt some pressure toward progressive reforms. There were a thousand signs of a turn.

And then what?

Much heralded within the movement for months, though largely ignored by the media, Occupy's "general strike" of May 1 came and went in good spirit, at least in New York, but without breaking new ground. In Manhattan, there were two big marches, some school walkouts, some arrests, and several small offshoots aimed at banks and such. All in all, the day felt like a regathering of the tribes, which, after a winter of dispossession and dispersal, was good for movement morale—certainly much better than a sparse turnout would have been.

May Day activity was reasonably general, at least in New York and the San Francisco Bay Area. It was not, however, much of a strike. At its peak, the exuberant New York crowd numbered some fifteen or twenty thousand—

up there with the big-turnout events of the fall—and was almost entirely peaceful, featuring floats, playful signs, parents with children, and unionists along with Occupy groups from the outer boroughs and veterans of Zuccotti Park. In the Bay Area, thousands of nurses did walk out, and workers persuaded Occupy Oakland not to go ahead with its original plan to block the Golden Gate Bridge, thus scoring a point for damage control. A few thousand marched through Oakland. But the night before, a rump march had trashed windows in the Mission District of San Francisco. Taken aback, Occupy San Francisco condemned the sideshow: "We consider these acts of vandalism and violence a brutal assault on our community and the 99%."

Was this the long-feared eruption of "Days of Rage" mayhem more concerned about "inflicting damage to the 1 percent" (as if broken windows would smite boardrooms everywhere) than winning new friends and influencing people? Scott Rossi, an observant Occupy San Francisco activist, posted a report saying "we were hijacked." The flavor of his comments is worth citing at length:

> You remember those asshole jock bullies in high school? Well that was who was leading the march tonight. Clean cut, athletic, commanding, gravitas not borne of charisma but of testosterone and intimidation. They were decked out in outfits typically attributed to those in the 'black bloc' spectrum of tactics, yet their clothes were too new, and something was just off about them. They were very combative and

nearly physically violent with the livestreamers on site, and got ignorant with me, a medic, when I intervened. . . . I didn't recognize any of these people. Their eyes were too angry, their mouths were too severe. They felt "military" if that makes sense. Something just wasn't right about them on too many levels. . . . This wasn't directed against corporations or big banks, with the exception of one single ATM I saw smashed. This was specifically directed against mom and pop shops, local boutiques and businesses, and cars.

Rossi was at pains to stick to the evidence of his senses:

I'm not one of those tin foil hat conspiracy theorists, I don't subscribe to those theories that Queen Elizabeth's Reptilian slave driver masters run the Fed. I've read up on agent provocateurs and plants and that sort of thing and I have to say that without a doubt, I believe 100% that the people that started tonight's events in the Mission were exactly that.

There were similar reports in New York.

Nevertheless, Occupiers took pleasure in seeing the American left join the rest of the world, where May Day, which originated as a response to a police massacre at an 1886 Chicago demonstration for the eight-hour day, is routinely celebrated by parties of the left as a workers' holiday. Some Occupy Direct Action regulars, like Amin Husain, took part in open-air classes on the history of May Day. It was worth knowing that President Grover Cleveland intervened in 1887 to see that America's workers would celebrate *their* Labor Day in September, not May. In 1921, during the post–World War I antisubversive panic, May 1 was designed Loyalty Day— perhaps the most obscure, least celebrated designation of any American holiday.

At a postmortem held by the Direct Action working group that supervised the day's actions, the mood was upbeat. The movement took pleasure in having emerged from its winter hibernation. For their part, the mainstream media, in predictable fashion, took a police-blotter approach to the day's activities. As is so often true—decades ago, I wrote about the tendency to cover demonstrations as if they were crime scenes—arrests were "the story." Reporters let the police set their agenda and frame their stories. None of the city's news organs showed the thousands clogging the streets, though pictures suggesting the scale of the turnout could easily have been taken from above—as some did on Facebook, in fact. No, the published photos showed dramatic confrontations between police and demonstrators. The *New York Daily News* offered its readers not any old shot of an arrest but a shot of a bug-eyed arrestee with a wispy white beard wearing a T-shirt that says either "DHARMA" or "KARMA" (the photo isn't clear). Typically, a *New York Times* online headline read: "Scores Cuffed or Cited by End of Day of Demonstrations."

For a sense of May Day's scale, one would have done better to turn to a *New York Post* editorial, which began "They're back—sort of" but graciously did mention "thousands" in its next paragraph:

> After hours in which the dregs of Occupy Wall Street largely failed in their vow to cause a day of "no work, no school" in New York, thousands of protesters made a mess of the evening commute for many folks by mobbing lower Manhattan. Terrific.

No surprise here: The *New York Post* was the *New York Post*, a platform for Rupert Murdoch's nasty, gloating, plutocracy-loving, union-hating, self-serving view of the world. But Murdoch's raw political prejudices could not frame the coverage outside his own wholly owned precincts, the *Post* and Fox News. How, then, to explain the more general belittlement, the reduction of events involving tens of thousands to the fringe activities of a few dozen getting arrested?

The media are jaded—occupationally so. Movements are messy, hard to find, and therefore hard to report, while arrests are easy to count—indeed, there are government agencies that do it for you. But journalistic skepticism goes further. By the unspoken code of the profession, to be jaded is proof of discernment, balance, and craft. Before May Day, the conventional judgment was that Occupy had essentially vanished. Now, it seemed to the nation's prime adjudicators of novelty, the movement was back—"sort of"—but its public manifestations were mostly old news. Old news was moldy news. Old news, in the end, was no news at all.

No news, anyway, if you still depended upon "old media" for your sense of what was "happening" in the world—that is, worth your attention. You might have been more or less oblivious to the Occupy movement when it erupted in the fall; you might have had mixed feelings, even sympathized. But one way or the other, six or seven months after the last news from Occupy, you would very likely have assumed that the movement had passed away, leaving barely a trace, or degenerated into those "mobs" that Eric Cantor had warned against.

But Occupy, of course, looked different if you were one of the stalwart core who dwelled mainly or exclusively in a movement world. In that case, the spirit of liberation you felt on May 1, even the joy at the way in which the morning's gray skies and rain yielded to afternoon sunshine, was what mattered. To you the actions felt gleeful and righteous at the same time. Eros had gone to work in the service of duty. The long, troubled winter—the wandering in the wilderness after Zuccotti Park, the fights over male supremacy and whether to sign a pledge of nonviolence, the breakdowns of General Assemblies, the chaos of Spokes Councils, the anger at sinister agents of co-optation taking orders from card-carrying members of the Democratic Party—could all be shoved back into the shadows. May Day was daylight. The movement was back, and to be back could seem to be itself the victory, never mind that the idea of a general strike was hyperbole and it was easy to observe that workers were still working and shoppers were still shopping. Never mind. There was undeniable joy in the spectacle of the return to the streets of lower Manhattan in large numbers. There was frolicking and heartfelt sentiment in the floats and the puppets. "We're all in the same boat," read the words painted onto a symbolic boat put together by the Lower East Side's Theater for the New City. There was a maypole, each ribbon inscribed with one of the accusations drawn from the September "Declaration of the Occupation of New York City" (see page 111). The ribbons were intended as a declaration that the issues of concern to Occupy's activists—however disparate they might appear to outside observers—were in fact connected. In that affirmation there would be, for old-timers like me, a distant echo. There was repeated here, in fact, an SDS slogan of the early sixties: *The issues are interconnected. They arise together and they need to be solved together.* That was an article of faith.

It wasn't only that the Occupy movement, like its radical precursor, the New Left, decades before, *imagined* these connections. It not only *thought* them; not only knew, in the back of its collective mind, that it could *explain* them (at least to anyone of reasonably good will who had time and patience enough to follow the argument). It felt in its bones that in its community, in its resolve, it *itself*—somehow, if inexplicably, perhaps only embryonically, in some trembling potentiality—embodied or prefigured the sort of ultimate solution that it so profoundly desired. It would be a solution in the sense of a melting together. It was liquefied. It flowed.

This was a declaration of identity and spirit, not strategy. The day of rejoicing itself felt like a large, even a grand achievement worth savoring. Having headed back to the streets, the prideful movement had triumphed over the grim counsels of what Freud called the reality principle. And yet in its beating heart it harked back to something larger than any pleasure principle, too. It felt, and acted upon, a call of what some in the cheerful crowds would have been embarrassed to call the spirit. Though many in the movement would not have welcomed the analogy, this was comparable to the way certain Christian congregations might have monitored their own sentiment and felt themselves transfigured or even holy. You could tell the inner movement that this was a self-enclosed sort of victory, and some of them would admit that such satisfactions did not radiate very far outward, that they rattled around in the confines of the movement's own echo chamber. They resolved to do more outreach and went back to planning for their next rally.

Moving deeper into May, as one movement event succeeded another, from disappointment to epiphany and from one epiphany to (or so one could always hope) another, the strategic questions remained unanswered—in fact, largely unasked. If you were not a hard core Occupier, then the May 1 revival only kicked down the road some awkward questions: What to do besides more of the same? Was the movement to be content with the hope that larger numbers would take to the streets next time, and if not, what was the alternative? The week after May Day, for example, brought a May 9 mobilization at the Bank of America stockholders meeting in

Charlotte, where some six hundred Occupiers—but evidently not the anticipated thousand or more—rallied and frolicked, and some shareholders inside the meeting leveled charges about the bank's handling of its loans. One shareholder later said, "No person in that meeting who stood and spoke to [the Bank of America board] had anything positive to say about their practices. Every voice that spoke was in dissent." Still, 92 percent of the shares present supported the CEO's $7 million pay package.

Where, then, was the leverage for tangible change? How would people's practical lives benefit from the movement's playful mobilizations? If the movement was to be wholly indifferent to concrete reforms, if it believed that finance capitalism or corporate capitalism or crony capitalism or neoliberalism was an indivisible black box, a take-it-or-leave-it monolith, then sooner or later the movement would be beating its head against the wall. There would be movement people who would dedicate months, even years, toward that end, but they would number a few tens of thousands, initially, and then dwindling numbers of thousands. And then?

Meanwhile, the country was in the thick of a presidential campaign that within a few months would either ratify the politics of austerity—the decades-long, largely victorious struggle of the plutocracy against the proverbial 99 percent—or resist it. Most Americans, however disillusioned they might be by political rituals, however revolted by the shabbiness and duplicity of what passes for political debate in the party system, would label it the central phenomenon of the year. The election would pit a cautious, thwarted, disappointing, sometimes progressive, sometimes populist president against a genuine (and wildly mendacious) plutocrat utterly beholden to his class, the veritable 0.1 percent, never mind the whole 1 percent. The result would either reinforce the oligarch-loving zeal of a Republican-majority Supreme Court or stand a chance of arresting it. It would encourage defenders of science or their fundamentalist enemies. It would, not least, elect the next House of Representatives, and one-third of the next Senate, bodies which would either consider reform legislation or bury it. Yet, for the most part, the core of Occupy could not be bothered. It was otherwise occupied. It was wrapped up in something else. It was wrapped up in itself.

As presently constituted, the most the core could do would still be no small thing. It could devise projects to welcome newcomers, widen the circles of citizens it could call on. With the support of an outer movement that includes millions, at least on paper, it could press politicians to open criminal investigations of the institutions that cut corners for huge profits' sake while undermining the stability of the global finance system, and to investigate the banks' shoddy foreclosure practices. It could overcome its aversion to demands, support progressive taxation, curb the banks' casino operations in earnest, and press to reduce loans to fair market value. It could devise decision-making networks that feel legitimate enough to convince thousands of activists that constructive work can get done even without the entire former population of Zuccotti Park. Meanwhile, an alliance of inner and outer movements could draft a charter of reforms and demand that politicians sign on.

At worst, the movement could spin its wheels until, lacking traction, they come off.

My emphasis in the third part of this book has been on the political outlook for Occupy. But I wish to conclude on a somewhat different note, which—not to put too fine a point on it—has to do with the movement's right to exist. It is a question of the intellectual, historical, and legal standing of the First Amendment right of the people "peaceably to assemble, and to petition the Government for redress of grievances." For beyond questions of strategy and tactics, on which Occupy supporters may legitimately disagree, there should be easy agreement that authorities all over have demonstrated a flagrant disrespect for the movement's right to assemble—which means disrespect for the core of the movement's identity and thus its right to exist as a legitimate component of popular self-government. For without public inspection or adequate legal challenge, let alone legal rationale, local governments have been extinguishing this right. This ought not to be allowed to continue. It certainly ought not to continue without a fight.

This is not the place, nor am I the authority, to make a thoroughgoing case on constitutional, legal, and historical grounds. That would be, as they say, above my pay grade. But briefly I would like to suggest a different approach to the matter and to commend the argument and the evidence to citizens who care about the liberties that undergird America's opportunity to govern itself in a manner worthy of the words "democracy" and "republic."

Those who have been paying attention know that it has become routine for police to disperse Occupy encampments, to confine demonstrators inside metal fences, corral them in plastic, and sequester them in "free speech zones" far removed from gatherings they want to influence, or denounce, or otherwise communicate with or about. Gradually it has come to pass that public spaces are treated as if they belong to the government, to be doled out by the spoonful, and not to the people, even though the First Amendment is quite explicit that what is forbidden is "abridging . . . the right of the people peaceably to assemble, and to petition the Government for a redress of grievances."

Originalists, please note: In 1791, the right to assemble was considered important enough to include in the first, foundational supplement to the Constitution, which was deemed essential to winning the ratification of states where anti-Federalist currents ran strong. The rights of the First Amendment—religion, speech, press, assembly, petition—were clumped together at the top of the Bill of Rights because they spoke of fundamentals that had been exercised by, and secured through, an American Revolution of not-so-distant memory.

Today, through a long-running bloat in police power

that ought to offend conservatives who are often rightly suspicious of government, freedom of assembly is, de facto, embattled—moribund as a matter of law though vigorously enacted on the ground by Occupy Wall Street. Occupy assembles—that's what it does. Its people *consult*. They discuss. They argue. They symbolize. They represent themselves.

I have maintained throughout this book that Occupy's distinctive feature has been its way of merging the powers of social media and face-to-face community, where the crowd overcomes its loneliness and publics come into their own. Yet today local governments presume assemblies guilty until proven innocent. This is precisely the reverse of what the First Amendment implies—and, as we shall see, what most state constitutions do as well. James Madison put the principle bluntly in a speech to the House of Representatives in 1794: it was in "the nature of republican government" that "the censorial power is in the people over the government, and not in the government over the people."

It would seem unlikely that the official, rampant disrespect shown to Occupy assemblies coast to coast stems from decisions made independently by hundreds of police officials. I mentioned above, on page 40, a conference call that linked worried mayors who wanted to consult about these damnable assemblies. More evidence has also come to light. One interurban call was characterized as about "general information-sharing and best practices," another officially downplayed as "like a therapy session." But when one considers the formidable armament deployed by police at Occupy demonstrations, along with stringent ordinances passed in Charlotte and Chicago (see page 198), it's clear that in present-day America, the right to assemble is being treated as a nuisance—not a cornerstone of the republic. On this subject, politicians, even liberals, remain mute. Civil liberties don't win many votes.

Even as the authorities wave it away, the right of assembly is the subject of very little discussion and few judicial decisions. Its intellectual standing has for years not been much probed. But fortunately, exceptions crop up among recent writings by three law professors. "In the past thirty years," writes John D. Inazu of Washington University law school in *Liberty's Refuge: The Forgotten Freedom of Assembly*, "the freedom of assembly has become little more than a historical footnote in American law and political theory." Both Inazu's book and another, *Speech Out of Doors: Preserving First Amendment Liberties in Public Places*, by Timothy Zick of William & Mary, go some considerable distance toward laying an intellectual foundation for elevating freedom of assembly to the stature it deserves. Zick in particular writes compellingly about the importance of an expressive conception of place. If a group of citizens wish to make a statement to or about Wall Street, then the place where they do so matters for the integrity of their case. There are only a limited number of public places where they may convincingly do that—and Zuccotti Park is one of them. And so there ought to be a very high bar before any government agency is permitted to stem such expression. Still another law professor, Ronald J. Krotoszynski Jr., of the University of Alabama, has published a book, *Reclaiming the Petition Clause: Seditious Libel, "Offensive" Protest, and the Right to Petition the Government for a Redress of Grievances*, that focuses on the right of petition, which he traces to an ancient right to confront the royal authorities face-to-face, not at a great remove, behind fences, sequestered in remote, forbidding cages designated, in sinister euphemism, "free speech zones."

There is another interesting feature to the right of assembly deserving of much attention and exploration, not least by conservatives who prefer to commend questions of fundamental liberty to state as opposed to federal authorities. *A preliminary search of state constitutions shows that the right to assembly is protected in almost all.* Pennsylvania seems to have set the pace in 1776:

> The citizens have a right in a peaceable manner to assemble together for their common good, and to apply to those invested with the powers of government for redress of grievances or other proper purposes by petition, address or remonstrance.

Connecticut adopted a text that differs in only a single word: it subtracts the word "together." Several other states stuck with "common good" and added "consult," as in this from New Jersey:

The people have the right freely to assemble together, to consult for the common good, to make known their opinions to their representatives, and to petition for redress of grievances.

Illinois has an almost identical version, and the talkier Massachusetts has this:

The people have a right, in an orderly and peaceable manner, to assemble to consult upon the common good; give instructions to their representatives, and to request of the legislative body, by the way of addresses, petitions, or remonstrances, redress of the wrongs done them, and of the grievances they suffer.

California, land of modernity and pepper spray, compresses the guarantee but retains "consult" and "common good":

The people have the right to instruct their representatives, petition government for redress of grievances, and assemble freely to consult for the common good.

Maryland declines the word "assembly" but affirms "[t]hat every man hath a right to petition the Legislature for the redress of grievances in a peaceable and orderly manner."

In Zuccotti Park's home state, New York, the state

constitution guarantees "the right to assemble and petition," as follows:

> No law shall be passed abridging the rights of the people peaceably to assemble and to petition the government, or any department thereof . . .

Weirdly, though, it groups the right of assembly in a section that goes on to speak to "divorce; lotteries; pool-selling and gambling; laws to prevent; pari-mutual betting on horse races permitted; games of chance, bingo or lotto authorized under certain restrictions."

If the states have the juridical standing that conservatives claim for them, where is their outcry when the right to assemble is treated as a tentative, revocable loan by the authorities? If liberals take liberties seriously, where is their own outcry? Where is the public good to be debated by citizens whose own good is at stake?

When I was young, I owned a tattered paperback copy of a 1954 book by a CBS radio newsman named Elmer Davis. Davis had crusaded against Senator Joseph McCarthy, and I liked his flat Indiana voice, which I remembered from the days when one listened to commercial radio for enlightenment. The book consisted of the texts of Davis's radio and newspaper commentaries, and it was called *But We Were Born Free*. I remember nothing about it in particular, but it stirred me, so much so that in my memory—but *only* there—the title comes with an exclamation point. But we were born free!

As I think about the authorities of this semi-democracy

who cavalierly junk the right to assemble because it is, for some people, sometimes, inconvenient—and when I think, for that matter, of much of the populace who side with this distaste and think that nothing is more important than private citizens going about their business oblivious of the common good—those words come back to mind: But we were born free!

May 10, 2012

ZUCCOTTI PARK:
A PHOTOGRAPHIC RECORD
By Victoria Schultz

ACKNOWLEDGMENTS

To write a book about a movement a few weeks old, whose headquarters were on a stone slab, possessed of no known structure or leadership, was the idea of my devoted and indefatigable agent, Ellen Levine. It seemed like a crazy idea, so I said: Sure.

At HarperCollins, Cal Morgan saw the possibilities of an adventure in publishing: a book that would be published the way Occupy was organized, by staking out some new territory in a hurry—in this case, an e-book to be ready by spring—and then seeing what developed and trying to keep up. Denise Oswald, my editor, got it. They left me alone when that was called for, and offered improvements when *that* was called for. Publishing a book should always be such a joy.

Travis Mushett was a splendid research assistant.

Liel Leibovitz, Brian Morton, and my beloved wife, Laurel Cook, read the manuscript in various phases and gave me nicely concocted mixtures of encouragement and critique. Hudson Kidwell, age six, offered his own frequent encouragements, although I didn't follow his suggestion for a title: *Pop Pop's Investigation of Occupied Wall Street*. My students and colleagues were kind and forbearing.

A lot of people helped in a lot of ways—advice, facts, address books, schmoozing. I relished and learned from conversations with and contacts from Anthony Barnett, Gwenda Blair, Alexandra Boonekamp, Will Bunch, Pablo Calvi, Marianne Debouzy, Jim Dingeman, Rodrigo Dorfman, Marshall Ganz, Danny Goldberg, David Greenberg, Michael Greenberg, John Heilemann, Michael Hirsch, John Hulsey, Marisa Jahn, Michael Janeway, Marilyn Katz, Sarah Leonard, Michael Levitin, Susan Ossman, Jon Phoenix, Srdja Popovic, Don Rose, Ruth Rosen, Rachel Rosenfelt, Carne Ross, Julian Rubenstein, Danny Schechter, Matthew Taylor, Richard Yeselson, and quite a few others whose names I never got, as well as the many journalists, credited in notes, who did plenty of leg-, brain-, and notebook-work.

Pablo Benson, Max Berger, Mark Bray, Nicole Carty, David DeGraw, Bill Dobbs, Michael Fix, Marty Frates, Yoni Golijov, Adam Green, David Haack, Matt Browner Hamlin, Amin Husain, Brooke Lehman, Will Jesse, Jennifer Lightfoot, Meaghan Linick, Yotam Marom, Leonardo Mejia, Ed Needham, Tanya Paton, Sean Patrick,

Rob Robinson, Justin Ruben, Matt Smucker, Eliot Tarver, and Shen Tong sat still for interviews. I thank them for their generosity and patience.

We elders need dynamic youth. It's my profound wish to have given something back.

NOTES

Part One: Two Energy Centers

1. *Pioneers*

"Something has been": Interview by Naomi Klein, *The Nation,* Jan. 9, 2012, http://www.thenation.com/article/165530/why-now-whats-next-naomi-klein-and-yotam-marom-conversation-about-occupy-wall-street?page=full, accessed Jan. 15, 2012.

"the combination of hi-tech networking": Anthony Barnett, "The Long and Quick of Revolution," *openDemocracy,* Dec. 16, 2011, http://www.opendemocracy.net/anthony-barnett/long-and-quick-of-revolution, accessed Dec. 22, 2011.

"reality TV on steroids": The phrase comes from a character in Kevin Breslin's documentary "#WhileWeWatch" (http://www.whilewewatch.com).

Barrie Thorne: Barrie Thorne, "Resisting the Draft: An Ethnography of the Draft Resistance Movement," unpublished Ph.D. dissertation, Brandeis University, 1971, pp. 341–42.

What will you do: Germaine Greer, *The Female Eunuch* (London: Granada Publishing Limited, 1970), p. 331.

paneled, extravagant yachts: The point about the symbolic power of the yacht that floats free of national boundaries is made by Anthony Barnett in "Long and Quick" (note 2).

animal spirits: John Maynard Keynes, *The General Theory of Employment, Interest and Money* (New York: Macmillan, 1973 [1936]), pp. 149–50, 161–62.

go-go years: John Brooks, *The Go-Go Years: The Drama and Crashing Finale of Wall Street's Bullish 60s* (New York: Wiley, 1999 [1973]).

business civilization: James Truslow Adams, *Our Business Civilization: Some Aspects of American Culture* (New York: A. & C. Boni, 1929).

"Dear Americans": Jeff Sharlet, "Inside Occupy Wall Street," *Rolling Stone,* Nov. 24, 2011, http://www.rollingstone.com/politics/news/occupy-wall-street-welcome-to-the-occupation-20111110#ixzz1hUViYsjG, accessed Dec. 24, 2011.

movement in history: I owe the lapidary moment/movement contrast to Marshall Ganz, who used it at a Harvard Kennedy School of Government forum on Oct. 13, 2011. Subsequently the meme seems to have taken on a life of its own.

rebooting history: Quoted in Sharlet, "Inside Occupy Wall Street."

99 percent movement: DeGraw published his six-part manifesto, using the phrase "the 99 percent," in February 2010: "The Economic Elite Vs. The People of the United States of America," http://ampedstatus.org/full-report-the-economic-elite-vs-the-people-of-the-united-states-of-america, viewed Jan. 16, 2012.

"we hear it every day": "ON MAY 12, MAKE BIG BANKS AND MILLIONAIRES PAY," http://www.onmay12.org/action/call-to-action, accessed Dec. 9, 2011.

on June 14: http://nocutsny.wordpress.com/bloombergville -info, accessed Dec. 29, 2011. If we need any reminder that the scale of movement events is hard to predict, David A. Chen's *New York Times* report, "In 'Bloombergville,' Budget Protesters Sleep In," began: "it may not be quite on the scale of the so-called Walkerville sit-in in Wisconsin, or the protests in Tahrir Square in Cairo." http://cityroom.blogs.nytimes .com/2011/06/15/in-bloombergville-budget-protesters-sleep-in, accessed Dec. 14, 2011.

After two weeks: John Tarleton, "Bloombergville Lives: Consigned to a Sidewalk across from City Hall, Protesters Refuse to Give Up but Struggle to Increase Their Numbers," http://www.indypendent.org/2011/06/20/bloombergville, and "Bloombergville 13 Released, Reflect on a Night in Jail," http://www.indypendent.org/2011/06/30/bloombergville-13-re leased, accessed Dec. 29, 2011. Thanks to Yoni Golijov (personal communication, Dec. 27, 2011) for clarifying details about the spring and summer actions.

radical homesteaders: Will Bunch, *October 1, 2011: The Battle of the Brooklyn Bridge,* Kindle E-book, Location 88.

Isham Christie: Isham Christie, "Possibility, Universality, & Radicality: A Universal Chorus for Emancipation," *Tidal: Occupy Theory, Occupy Strategy,* No. 1, December 2011, http:// occupytheory.org/TIDAL_occupytheory.pdf, accessed Dec. 26, 2011; Christie, interview on Free Speech TV, Sept. 22, 2011, http://blip.tv/free-speech-tv/newswire-skype-092212v2-5578874, accessed Dec. 26, 2011; http://sds.averysmallbird.com/?p=33, accessed Dec. 21, 2011; http://www.warresisters.org/staff, accessed Dec. 26, 2011.

in 1770: Peter Maass, "Toppling Dictators in the YouTube Age," New Yorker online, April 12, 2011, www.newyorker.com/ online/blogs/newsdesk/2011/04/toppling-dictator-movements .html, accessed March 11, 2012.

Privately Owned Public Space: http://www.nyc.gov/html/
dcp/html/priv/priv.shtml, accessed Nov. 25, 2011.

against the 1 percent: The anthropologist David Graeber
has been credited with "We are the 99 percent," while the
blogger David DeGraw had started using "the 99 percent" in
2010. DeGraw's account is "A Report from the Frontlines: The
Long Road to #OccupyWallStreet and the Origins of the
99% Movement," http://ampedstatus.org/a-report-from-the-front
lines-the-long-road-to-occupywallstreet-and-the-origins-of-the
-99-movement/, Sept. 29, 2011, accessed Jan. 16, 2012. (I in-
terviewed DeGraw by phone on Jan. 10, 2012.) It appears on a
flyer as early as August 9, more than a month before the occupation of Zuccotti Park. But more important than a phrase's
precise origins is the evident fact that it caught on. Lots of
slogans show up, but few are chosen.

May issue of Vanity Fair: Joseph E. Stiglitz, "Of the 1%, by
the 1%, for the 1%," *Vanity Fair,* May 2011, http://www.vanity
fair.com/society/features/2011/05/top-one-percent-201105, accessed Dec. 29, 2011.

By one calculation: Dave Gilson, "Charts: Who Are the
1 Percent?" *Mother Jones,* Oct. 10, 2011, http://motherjones
.com/mojo/2011/10/one-percent-income-inequality-OWS, accessed Jan. 7, 2012. The very richest skewed more toward executives. According to a study cited by Paul Krugman ("We
Are the 99.9 %" *New York Times,* Nov. 24, 2011, http://www
.nytimes.com/2011/11/25/opinion/we-are-the-99-9.html, accessed Jan. 7, 2012), 43 percent of the top 0.1 percent were
nonfinancial executives, 18 percent were in finance, and
12 percent were lawyers or involved in real estate.

"goddamned hippies": Kevin Sheehan and Rebecca Harsh-
barger attributed the phrase to "one irate driver" in "Wall St.
protestors shut down Brooklyn Bridge, *New York Post*, Oct. 1,
2011, http://www.nypost.com/p/news/local/brooklyn/wall_st_pro

testors_shut_down_brooklyn_yEqcq6EsqgJ1cSy0rXhkzJ#ixzz1
ZZenEpUn, accessed Feb. 8, 2012.

Will Bunch: Bunch, *October 1, 2011,* Location 701.

the Times *ran:* Colin Moynihan, "Protesters Find Wall Street
Off Limits," *New York Times,* Sept. 18, 2011, p. A22.

Debbs says: Telephone interview, Jan. 14, 2012.

Newark Star–Ledger: "Occupy Wall Street demonstrators send message: Where are the jobs?" http://blog.nj.com/
njv_editorial_page/2011/09/occupy_wall_street_demonstrato
.html, accessed Jan. 17, 2012.

live stream video feed: www.livestream.com/globalrevolu
tion, accessed Jan. 17, 2012.

Max Berger: Quoted in John Heilemann, "2012 = 1968?"
New York, Nov. 27, 2011, http://nymag.com/print/?/news/poli
tics/occupy-wall-street-2011-12/, accessed Jan. 17, 2012.

one cop even: Sharlet, "Inside Occupy Wall Street."

Occupy Your Life: "5 Things You Can Do Now," *Occupied
Wall Street Journal,* No. 5, November 2011, p. 4.

Jesse LaGreca: Sharlet, "Inside Occupy Wall Street."
LaGreca became famous for an articulate rant to a Fox News
reporter, Oct. 2, that somehow never aired but went viral courtesy of an independent video uploaded to YouTube (http://
www.youtube.com/watch?feature=player_embedded&v=6yrT
-0Xbrn4, accessed Dec. 24, 2011). For his statement on voting, see http://www.dailykos.com/story/2011/12/20/1047443/
-Occupy-Wall-Street), accessed Dec. 24, 2011.

numerical estimates: Interview, Shen Tong, Jan. 20, 2012;
Shen Tong with Marianne Yen, *Almost a Revolution* (Boston:
Houghton Mifflin, 1990).

New York Times *reporter:* Kevin Roose, "A Blow to Pinstripe
Aspirations," NYT, Nov. 22, 2011, http://dealbook.nytimes
.com/2011/11/21/wall-st-layoffs-take-heavy-toll-on-younger-wor
kers/, accessed Nov. 24, 2011.

"ragtag group": Nelson D. Schwartz and Eric Dash, "In Private, Wall St. Bankers Dismiss Protesters as Unsophisticated," *New York Times,* Oct. 15, 2011, p. B1.

2. *Surprise! Popularity and Polarization*

"Genia Bellafante's": http://www.nytimes.com/2011/09/25/nyregion/protesters-are-gunning-for-wall-street-with-faulty-aim.html, accessed Dec. 8, 2011.

NYPD Deputy Inspector: "USLaw.com's Analysis of Occupy Wall Street Participants Pepper Sprayed by New York Police Department," http://www.uslaw.com/occupywallstreet, accessed Dec. 2, 2011; Bunch, *October 1, 2011,* Location 161.

support for withdrawal: The figure fluctuated wildly, depending at least in part on how the question was worded. Harris Surveys, Dec. 1964–Dec. 1965, retrieved Dec-6-2011 from the iPOLL Databank, The Roper Center for Public Opinion Research, University of Connecticut, http://www.ropercenter.uconn.edu.ezproxy.cul.columbia.edu/data_access/ipoll/ipoll.html.

in 1969: Harris Survey, Sep, 1969. Retrieved Dec-6-2011 from the iPOLL Databank, The Roper Center for Public Opinion Research, University of Connecticut, http://www.ropercenter.uconn.edu.ezproxy.cul.columbia.edu/data_access/ipoll/ipoll.html.

in 1970: Virginia Slims American Women's Poll 1970, Aug, 1970. Retrieved Dec-6-2011 from the iPOLL Databank, The Roper Center for Public Opinion Research, University of Connecticut, http://www.ropercenter.uconn.edu.ezproxy.cul.columbia.edu/data_access/ipoll/ipoll.html.

attitudes toward gays: Roper Commercial Survey, Oct, 1971. Retrieved Dec-6-2011 from the iPOLL Databank, The Roper

Center for Public Opinion Research, University of Connecticut http://www.ropercenter.uconn.edu.ezproxy.cul.columbia.edu/data_access/ipoll/ipoll.html.

industrial union drive: In May 1939, they asked: "Do you think workers should have the right to join together in a union, in order to bargain with their employers?" Seventy-nine percent said yes, thirteen percent no, seven percent didn't know. [USGALLUP.39-158.QA06] But earlier that year, seventy-five percent agreed that "every labor union should be required to take out a license (permit) from the Federal Government," [USGALLUP.MA0139.RB06], and in December, when asked, "Should an employer have the right to ask people applying for a job whether or not they belong to a union?" seventy percent answered Yes. [USGALLUP.39-179.QAB06A] In October 1937, asked whether "the attitude of the Roosevelt Administration toward union labor is too friendly or not friendly enough," thirty-nine percent said "too friendly," fifteen percent "not friendly enough," and forty-six percent "about right." [USGALLUP.37-102.Q01A] In July 1937, Gallup asked: "John L. Lewis is trying to organize the workers of the Ford Motor Company into a CIO (Congress of Industrial Organizations) labor union. Do you hope he succeeds?" Twenty-two percent said yes, 61 percent no, 17 percent didn't know. [USGALLUP.37-90.Q03] In December 1937, they asked: "In the present dispute between Henry Ford and the Automobile Workers Union are you [sic] sympathies with Ford or with the union?" Sixty-six percent were with Ford, 34 percent with the union. [USGALLUP.011938.R01] In two polls during the spring of 1937, about 1.8 times as many respondents preferred AFL-style craft unions to CIO-style industrial unions. [USGALLUP.37-76.Q05A, USGALLUP.JL0437.R04] Gallup began polling in 1935. All data courtesy of iPOLL.

supermajority support: See, for example, the polls cited by Bruce Bartlett, "Americans Support Higher Taxes. Really.,"

June 29, 2011, http://capitalgainsandgames.com/blog/bruce-bartlett/2292/americans-support-higher-taxes-really, accessed Nov. 29, 2011; and a CNN poll from Oct. 14–16, 2011, cited in Geoff Garin, "The New Focus on Income Inequality and the Impact on Democrat Prospects for 2012," Nov. 21, 2011, http://big.assets.huffingtonpost.com/IncomeInequlaityMemo .pdf, accessed Dec. 12, 2011.

nearly twice as many: Pew Research Center/*Washington Post* poll, Sept. 1–4, 2011, http://www.pollingreport.com/consumer2 .htm, accessed Dec. 12, 2011.

one November poll: ABC News/*Washington Post* poll released Nov. 16, 2011, http://www.langerresearch.com/ uploads/1129a5IncomeInequality.pdf, accessed Dec. 12, 2011.

A different November: Garin, "New Focus".

to evict Occupy activists: John Stoehr, "Occupy New Haven: The Last Camp Standing?" http://nationalmemo.com/article/ occupy-new-haven-last-camp-standing, accessed Dec. 27, 2011.

conference call beforehand: Allison Kilkenny, "Did Mayors, DHS Coordinate Occupy Attacks?" *In These Times,* Nov. 16, 2011, http://www.inthesetimes.com/uprising/entry/12303/may ors_dhs_coordinated_occupy_attacks, accessed Dec. 16, 2011.

Justice Department official: Rick Ellis, "Update: 'Occupy' crackdowns coordinated with federal law enforcement officials," Nov. 15, 2011, http://www.examiner.com/top-news-in-minne apolis/were-occupy-crackdowns-aided-by-federal-law-enforce ment-agencies, accessed Jan. 6, 2012.

"As we have learned": Aaron Bady, "The case for making a storm in the ports," *Salon.com,* Dec. 14, 2011, http://www.salon .com/2011/12/14/the_case_for_making_a_storm_in_the_ports/ singleton, accessed Dec. 16, 2011.

one poll showed: USA Today/Gallup Poll, Nov. 19–20, 2011, http://www.pollingreport.com/politics.htm, accessed Dec. 14, 2011.

In other polls: NBC News/*Wall Street Journal* Poll, Dec. 7–11, 2011, http://www.pollingreport.com/politics.htm, accessed Dec. 14, 2011.

Pew found: "Even those who say they agree with the concerns raised by the movement," wrote the Pew analysts, "are somewhat divided over its tactics: 50% approve, while 40% disapprove." http://www.people-press.org/2011/12/15/section-2-occupy-wall-street-and-inequality/, accessed Dec. 15, 2011.

blurred the tea leaves: Pew found 44 percent either "strongly" or "somewhat" supporting Occupy—almost double the roughly concurrent NBC News/*Wall Street Journal* figure of 24 percent, cited above—but Pew's 44 percent was the sum of 15 percent who said they "strongly supported" the movement and 29 percent who supported it "somewhat." NBC/WSJ hadn't offered the two options. So, in effect, Pew offered the opportunity to express trepidation without abandoning the Occupy cause altogether.

slight edge to Occupy: ABC News/*Washington Post* Poll (see note 17). Strikingly, one-third of Tea Party supporters also supported Occupy—and vice versa.

"Fight the people": Jeremy Varon, *Bringing the War Home: The Weather Underground, the Red Army Faction, and Revolutionary Violence in the Sixties and Seventies* (Berkeley: University of California Press, 2004), p. 139, Google eBook, accessed Feb. 8, 2012.

3. *Energy Is Eternal Delight*

Senator Charles E. Schumer: John Harwood, " 'Spreading the Wealth' in Democrats' Favor," *New York Times,* Nov. 28, 2011, http://thecaucus.blogs.nytimes.com/2011/11/27/spreading-the-wealth-in-democrats-favor, accessed Nov. 28, 2011.

Schumer's intake: Center for Responsive Politics, http://www.opensecrets.org/politicians/industries.php?cycle=2012&cid=n00001093&type=I&newmem=N, accessed Nov. 29, 2011.

Andrew Cuomo: "Cuomo hedges on millionaire's tax," Dec. 1, 2011, http://www.lohud.com/article/20111201/OPINION/312010009/Cuomo-hedges-millionaire-s-tax, accessed Dec. 14, 2011; Michael Gurley, "State hungrily eyes wealthy," *Albany Times-Union*, Dec. 10, 2011, http://www.timesunion.com/business/article/State-hungrily-eyes-wealthy-2394810.php, accessed Dec. 17, 2011.

Previously: Eric Alterman, "Governor Cuomo Is Still Governor One Percent," *The Nation,* January 9–16, 2012, http://www.thenation.com/article/165305/governor-cuomo-still-governor-one-percent, accessed Jan. 17, 2012.

viewed it unfavorably: New York Times/CBS Poll, Aug. 2–3, 2011, http://s3.documentcloud.org/documents/229918/the-full-results-from-the-new-york-times-and-cbs.pdf, accessed Nov. 29, 2011.

Tea Party largely: Theca Skocpol and Vanessa Williamson, *The Tea Party and the Remaking of Republican Conservatism* (New York: Oxford University Press, 2012).

local news reports: Peter Dreier and Todd Gitlin, "Demonstrations at CEO Mansions? Ho Hum," *Columbia Journalism Review* online, Oct. 5, 2009, http://www.cjr.org/behind_the_news/demonstrations_at_ceo_mansions.php?page=all, accessed Nov. 30, 2011.

Tea Party coverage: Nate Silver, "Police Clashes Spur Coverage of Wall Street Protests," *New York Times* online, Oct. 7, 2011, http://fivethirtyeight.blogs.nytimes.com/2011/10/07/police-clashes-spur-coverage-of-wall-street-protests, accessed Nov. 30, 2011. Silver consulted the NewsLibrary.com database, which monitors some 4,000 primarily American news organs.

Tea Party was credited: In fact, it has been cogently argued that the Tea Party was routinely *overcredited* with the Republican victories of November 2010. John R. Bond, Richard Fleisher, and Nathan Alderson, "Was the Tea Party Responsible for the Republican Victory in the 2010 House Elections?," paper prepared for the 2011 meeting of the American Political Science Association, Sept. 2011, http://papers.ssrn.com/sol3/papers.cfm?abstract_id=1912707, accessed Nov. 30, 2011.

Brooklyn Bridge: Al Baker, Colin Moynihan, and Sarah Maslin Nir, "Police Arrest More Than 700 Protesters on Brooklyn Bridge," *New York Times* online, Oct. 1, 2011, http://cityroom.blogs.nytimes.com/2011/10/01/police-arresting-protesters-on-brooklyn-bridge, accessed Dec. 2, 2011; Bunch, *October 1, 2011,* Location 469.

foreign press: Telephone interview, Bill Dobbs, Jan. 14, 2012.

Within seven weeks: Dylan Byers, "Occupy Wall Street is winning," Nov. 11, 2011, http://www.politico.com/blogs/bensmith/1111/Occupy_Wall_Street_is_winning.html, accessed Nov. 25, 2011.

order pizzas: "Pizza parlor keeps protesters happy with its 'Occupier' special," *The Gazette* (Colorado Springs), Oct. 5, 2011, http://www.gazette.com/articles/parlor-126207-pizza-happy.html, accessed Dec. 24, 2011.

John Pike: Nov. 18, 2011, http://www.youtube.com/watch?v=WmJmmnMkuEM, accessed Dec. 27, 2011, by which time it had been viewed more than 2.7 million times.

window-smashing: http://www.youtube.com/watch?v=86XhCwHhwn8, accessed Dec. 15, 2011.

Part Two: The Spirit of Occupy

4. *Oases and Bases*

Working Groups: http://www.nycga.net/groups/, accessed Jan. 29, 2012.

"Monty Python *sketch*": Interview, Shane Patrick, Jan. 3, 2011.

Some looked at: The first two are quoted by Justin Elliott, "New York Post Declares war on Occupy Wall Street," *Salon*, Nov. 4, 2011, http://www.salon.com/2011/11/04/new_york _post_declares_war_on_occupy_wall_street," the others are posted to Candice M. Giove and Kenneth Garger, "Exiting Occupy Home," *New York Post*, Jan. 29, 2012, http://www.nypost .com/p/news/local/exitingoccupy_home_TyLpb8eFa5vSYfmuv VwoCL, accessed Jan. 29, 2012.

"*Woodstock vibe*": "Teddy Mapes, 46, was working in construction as a pipefitter before he was lured into the movement by what he described as Zuccotti Park's 'Woodstock vibe.'" Quoted in Adam Kaufman, "Occupy Wall Street Plans To 'Occupy Christmas,'" *Huffington Post*, Dec. 21, 2011, http:// www.huffingtonpost.com/2011/12/21/occupy-wall-street-christ mas_n_1163869.html, accessed Dec. 22, 2011.

shit is fucked: http://www.flickr.com/photos/jimkiernan/622 4561169/, accessed Dec. 27, 2011.

making less in 2011: Ellen Ruppel Shell, "Is Work Still Meaningful?" *Atlantic*, Dec. 16, 2011, http://www.theatlantic.com/ business/archive/2011/12/is-work-still-meaningful/250131/, accessed Dec. 24, 2011.

the workforce: John Irons, "11 Telling Charts from 2011," http:// www.epi.org/publication/11-telling-charts-about-2011-economy, accessed Dec. 24, 2011.

"*Take me to your leader!*": *Time* offered this sample in 1958: "A Martian lands in Paris, spots Brigitte Bardot and, electric-bulb eyeball flashing furiously, demands: 'Take me to your leader—later.'" "Out of This World," Dec. 1, 1958, http://www.time.com/time/magazine/article/0,9171,891982,00.html, accessed Dec. 24, 2011.

"Jews Control Wall Street": Michelle Goldberg, "One Percent," *Tablet,* Oct. 18, 2011, http://www.tabletmag.com/news-and-politics/80922/one-percent, accessed Dec. 24, 2011. The photo illustrating Goldberg's article shows a man holding a sign that reads: "Google: Zionists control Wall St." while a woman standing next to him holds one reading "WHO IS PAYING THIS GUY?" and another declares that he "doesn't represent OWS."

Christmas Eve: http://communityrelations.nycga.net/2011/12/24/occupygiftboxes, accessed Dec. 24, 2011.

relating to itself: Meredith Hoffman, "Protesters Debate What Demands, if Any, to Make," *New York Times,* Oct. 16, 2011, http://www.nytimes.com/2011/10/17/nyregion/occupy-wall-street-trying-to-settle-on-demands.html, accessed Dec. 24, 2011.

Think Tank: New School, Occupy Onwards conference, Dec. 18, 2011.

even crazy: That was the word used by a delicatessen counterman, Latino, across the street from Zuccotti Park when I asked his opinion on Oct. 5.

conviviality: Anthony Barnett described the May atmosphere in Madrid's Puerto del Sol this way: "It is both familiar and strange . . . like entering an eastern bazaar or souk. It creates the same dreamlike suddenness of going from outside into an enclosed but also public space: intensely busy, crowded with its own rhythm, that assaults your senses." "The Long and Quick of Revolution," *openDemocracy.net,* Dec. 16, 2011, http://www.opendemocracy.net/anthony-barnett/long-and-quick-of-revolution, accessed Dec. 22, 2011.

A *few of the campers:* Jonathan Matthew Smucker, "The Tactic of Occupation & the Movement of the 99%," Nov. 10, 2011, http://beyondthechoir.org/diary/117/the-tactic-of-occupation-the-movement-of-the-99, accessed Dec. 22, 2011.

5. *Rituals of Participation*

"The Occupy Movement": Brooke Lehman (writing as Brook Muse), "From GA to Spokes Council," *Occupy! An OWS-Inspired Gazette,* #2, p. 9.

"Others cautioned": See, for example, jemmcgloin at http://groups.google.com/group/september17/browse_thread/thread/fb033d10ed22311a#, accessed Jan. 7, 2012.

By December 8: Christopher Chase-Dunn and Michaela Curran-Strange, http://newsroom.ucr.edu/2813, accessed Dec. 27, 2011.

"I inherited money": http://www.paymytaxbill.com/wp-content/uploads/2011/10/one-percent-tax-me.jpg, accessed Dec. 29, 2011.

Bat-Signal: http://www.youtube.com/watch?v=vTX_8D7lDxU, accessed Dec. 29, 2011.

a Russian dissident: Glenn Kates, "Russians Rally for Activist in Hopes of Inspiring a Movement," *New York Times,* Dec. 30, 2011, http://www.nytimes.com/2011/12/30/world/europe/russians-rally-for-sergei-udaltsov.html, accessed Dec. 30, 2011.

in love with itself: Here I paraphrase the uneven but frequently insightful Slovenian philosopher Slavoj Zizek, "Don't Fall in Love with Yourselves," in Astra Taylor, et al., eds, *Occupy! Scenes from Occupied America* (New York: Verso, 2011), p. 66.

"an antidote": Interview, Brooke Lehman, Dec. 28, 2011, and Lehman (writing as Brook Muse), "From GA to Spokes Council," p. 9.

facilitators in Occupy: Telephone interview, Mark Bray, Jan. 13, 2012.

"empathy and solidarity": Lehman (writing as Brook Muse), "From GA to Spokes Council," p. 9.

insanely accelerated culture: Richard Kim, "We Are All Human Microphones Now," Oct. 3, 2011, http://www.thenation.com/blog/163767/we-are-all-human-microphones-now, accessed Dec. 19, 2011.

"we go apoplectic": Una Spenser, "#occupywallstreet: a primer on consensus and the General Assembly," Oct. 8, 2011, http://www.dailykos.com/story/2011/10/08/1022710/---occupywallstreet:-a-primer-on-consensus-and-the-General-Assembly, accessed Dec. 23, 2011.

"almost like a chair": Quoted in Jeff Sharlet, "Inside Occupy Wall Street," *Rolling Stone,* Nov. 8, 2011, http://www.rollingstone.com/politics/news/occupy-wall-street-welcome-to-the-occupation-20111110?page=2, accessed Dec. 24, 2011.

"worse than Congress": Interview, Jan. 15, 2012.

Berger, twenty–six: Interview, Jan. 15, 2012, and Max Berger, "What If the Occupy Movement is a Revolution," *Huffington Post,* Nov. 2, 2011, http://www.huffingtonpost.com/max-berger/what-if-the-occupy-moveme_b_1072174.html, accessed Jan. 18, 2012.

6. *The Evolution of Horizontalism*

A kind of anarchism: The following five paragraphs draw on Todd Gitlin, "The Left Declares Its Independence," *New York Times,* Oct. 9, 2011, p. SR4, http://www.nytimes.com/2011/10/09/opinion/sunday/occupy-wall-street-and-the-tea-party.html?pagewanted=all, accessed Dec. 19, 2011.

Jo Freeman: Jo Freeman, "The Tyranny of Structurelessness," first published 1972, revised version at http://www.jofreeman.com/joreen/tyranny.htm, accessed Dec. 20, 2011.

Argentine radicals: Marina Sitrin, ed., *Horizontalism: Voices of Popular Power in Argentina* (Oakland: AK Press, 2006). Marcela Valente, "Worker-run Companies Quietly Surviving," Nov. 8, 2010, http://ipsnews.net/news.asp?idnews=53488, accessed Jan. 17, 2012. For an early account of the founding of worker-owned cooperatives, see Pablo Calvi, "Modelo de crisis: las empresas que nacen sin empresarios," Clarín, Oct. 14, 2001, http://edant.clarin.com/suplementos/economico/2001/10/14/n-00411.htm, accessed Jan. 17, 2012.

some radicals warned: Hilary Wainwright, "European Social Forum: debating the challenges for its future," Transnational Institute, 2004, http://www.tni.org//archives/act/16321, accessed Jan. 15, 2012.

"take the bull": http://occupywallst.org/article/august_2nd_wall_street_assembly, accessed Dec. 23, 2011; http://takethesquare.net/wp-content/uploads/2011/07/occupyr1.jpg, accessed Dec. 26, 2011; Drake Bennett, "David Graeber, the Anti-Leader of Occupy Wall Street," *Bloomberg Businessweek,* Oct. 26, 2011, http://www.businessweek.com/magazine/david-graeber-the-antileader-of-occupy-wall-street-10262011_page_2.html, accessed Dec. 20, 2011.

A Greek anarchist: Personal communication, Yoni Golijov, Dec. 27, 2011, and Sharlet, "Inside Occupy Wall Street."

Alexis de Tocqueville's: "Americans of all ages, all conditions, and all dispositions constantly form associations. . . . I have often admired the extreme skill with which the inhabitants of the United States succeed in proposing a common object for the exertions of a great many men and in inducing them voluntarily to pursue it." Alexis de Tocqueville, *Democracy in America,* trans. Henry Reeve, Book II, Chap. 5, http://

xroads.virginia.edu/~HYPER/DETOC/ch2_05.htm, accessed Dec. 19, 2011.

modeling universal health care: Interview, Jennifer Lightfoot, Dec. 1, 2011.

David the Medic: http://www.nycga.net/members/daviddamedic/activity/59936, accessed Dec. 19, 2011.

Chabad, among other groups: http://www.youtube.com/watch?v=ATjkY0At96c&feature=related, accessed Dec. 19, 2011.

one drummer protested: "Drumming and the Occupation," Oct. 24, 2011, http://occupywallst.org/article/drumming-and-occupation, accessed Dec. 19, 2011; Sharlet, "Inside Occupy Wall Street."

Blair Mountain: Mark Johanson, "Saving Blair Mountain: Hundreds March in West Virginia," *International Business Times,* June 7, 2011, http://www.ibtimes.com/articles/158764/20110607/blair-mountain-west-virginia-march-protest-environmental-action-strip-mining-west-virginia-appalachi.htm, accessed Dec. 23, 2011.

Newt Gingrich: Nia-Malika Henderson, "Gingrich to Occupy protesters: 'Take a bath,'" *Washington Post,* Nov. 21, 2011, http://www.washingtonpost.com/blogs/election-2012/post/gingrich-to-occupy-protesters-take-a-bath/2011/11/21/gIQAqbhBiN_blog.html, accessed Dec. 19, 2011.

Anthony Barnett: Anthony Barnett, "The Long and the Quick of Revolution," Dec. 16, 2011, http://www.opendemocracy.net/anthony-barnett/long-and-quick-of-revolution, accessed Dec. 27, 2011.

7. Splendors and Miseries of Structurelessness

one fervent supporter: Brook Muse, "From GA to Spokes Council," p. 9.

People of Color: Audrea Lim, "Love Affair," *Occupy! An OWS-Inspired Gazette,* #2, p. 37.

Sean McKeown: http://www.nycga.net/members/smckeown/activity/64079, accessed Dec. 23, 2011.

spokes council: Rosie Gray, "Occupy Wall Street Debuts the New Spokes Council," Nov. 8, 2011, http://blogs.village voice.com/runninscared/2011/11/occupy_wall_str_25.php, accessed Dec. 26, 2011.

financial decisions: Ben Berkowitz and Chris Francescani, "Occupy Wall Street finds money brings problems too," Reuters, Nov. 2, 2011, http://www.reuters.com/article/2011/11/02/us-usa-wallstreet-protests-money-idUSTRE7A12DY20111102, accessed Dec. 26, 2011.

a women's caucus: Melanie Butler, "Women Being Disrespected at Occupy Wall Street?" Nov. 14, 2011, http://www.opposingviews.com/i/money/recession/women-being-disrespected-occupy-wall-street, accessed Dec. 26, 2011.

"it was a shock": Interview, Brooke Lehman, Dec. 28, 2011. *Occupied Wall St Journal,* #5, Nov. 2011, p. 2.

8. And Leaderlessness

Occupy Denver: Occupied Wall St Journal, #5, Nov. 2011, p. 2.

"proposal so sound": Nikil Saval, "Scenes from Occupied Philadelphia," in Taylor et al., eds., *Occupy!,* p. 158.

celebrities emerge: For this point I am indebted to Travis Mushett.

Shamar Thomas: http://www.youtube.com/watch?feature

=player_embedded&v=N9HvJhilJzo, accessed Dec. 26, 2011, and http://inthesetimes.com/uprising/entry/12484/protesters _arrested_for_public_mic_check1/, accessed Jan. 4, 2011.

"facilitators as leaders": Leo Eisenstein, "Facilitation Situation," *Occupy!*, #2, p. 4.

9. *The Movement as Its Own Demand*

bipartisan: The congressional vote is broken down by party at http://en.wikipedia.org/wiki/File:Gramm-Leach-Bliley_Vote _1999.png, accessed Jan. 1, 2012.

In October: Justin Elliott, "Occupy Wall Street looks ahead to 2012," *Salon,* Nov. 11, 2011, http://www.salon .com/2011/11/11/occupy_wall_street_looks_ahead_to_2012/si ngleton, accessed Jan. 14, 2012.

"Freedom Movement": Conversation, Marshall Ganz, Oct. 13, 2011.

Citizens United: https://www.nycga.net/2012/01/03/nyc -general-assembly-minutes-132012/#more-7541, accessed Jan. 7, 2012. The Los Angeles city council had, a month earlier, unanimously passed a resolution to this effect, pithily holding that "the rights protected by the Constitution of the United States are the rights of natural persons only." Sarah Jones, "Los Angeles Passes Resolution to Call for an End to Corporate Personhood," *Politicussusa.com,* Dec. 7, 2011, http://www .politicususa.com/en/los-angeles-resolution-end-corporate-per sonhood, accessed Jan. 7, 2012.

Gabriel Willow: Quoted in Meredith Hoffman, "Protesters Debate What Demands, if Any, to Make," *New York Times,* Oct. 16, 2011, http://www.nytimes.com/2011/10/17/nyregion/ occupy-wall-street-trying-to-settle-on-demands.html, accessed Jan. 1, 2012.

99 Declaration: http://www.the-99-declaration.org, accessed Jan. 1, 2012.

on September 29: http://www.nycga.net/resources/declaration, accessed Jan. 1, 2012.

Jeff Sharlet: Sharlet, "Inside Occupy Wall Street."

10. *Wonders of Nonviolence*

Alex Vitale: Alex Vitale, "NYPD and OWS: A Clash of Styles," in Taylor, et al., eds., *Occupy!*, p. 75.

Max Berger: Interview, Jan. 15, 2012.

Linda Katehi: http://www.youtube.com/watch?v=8775ZmNGFY8, accessed Jan. 2, 2012.

police violence: http://wagingnonviolence.org/2011/11/calls-for-violence-once-again-turn-people-off, accessed Jan. 2, 2012.

stop the Draft Week: The Sixties: Years of Hope, Days of Rage (New York: Bantam, 1987), pp. 249–55.

on January 28: http://www.youtube.com/watch?v=38gIB-d2A3M; Beth Duff-Brown, "Occupy protest rekindles debate about flag-burning," Associated Press, Jan. 30, 2012, http://bit.ly/xeBEEj; Susie Cagle, "Occupy protests in Oakland and New York: a weekend of police clashes," http://www.guardian.co.uk/world/us-news-blog/2012/jan/30/occupy-oakland-new-york-clashes; Aaron Bady, "From the Outside, Trying to Look In: Occupy Oakland's #J28," http://wp.me/p3FjM-1uz, accessed Jan. 30, 2012.

National Lawyers Guild: "Police Violence Targets Occupy Oakland Demonstration," Jan. 30, 2012, http://www.nlgsf.org/news/view.php?id=174, accessed Jan. 31, 2012.

Beyond the Choir: http://beyondthechoir.org.

Nine days into: http://beyondthechoir.org/diary/99/occupy

-wall-street-small-convergence-of-a-radical-fringe, accessed Jan. 2, 2012.

Max Berger: Interview, Max Berger, Jan. 15, 2012.

January 31: "@OccupyWallStNYC," Jan. 31, 2012, Twitter, https://twitter.com/#!/OccupyWallStNYC/status/16457308782 6837504, accessed Feb. 8, 2012.

"the fire next time": Baldwin's essay was first published in the *New Yorker* and then as a book by Dial Press.

Ruckus Society: http://www.ruckus.org/section.php?id=70, accessed Jan. 3, 2011.

Milosevic: The best account of the anti-Milosevic movement, at least in English, is in Tina Rosenberg, *Join the Club: How Peer Pressure Can Transform the World* (New York: Norton, 2011).

CANVAS workshops: An excerpt from my 2011 interview with Srdja Popovic of CANVAS is at http://newsmotion .org/blogs/newsmotion/nonviolent-resistance-todd-gitlin-and-sr dja-popovic, accessed Jan. 3, 2012.

Weathermen: Interview, Jim Dingeman, Jan. 8, 2012. See my *The Sixties,* pp. 393–94, and William Burr and Jeffrey Kimball, eds., "Nixon White House Considered Nuclear Options Against North Vietnam, Declassified Documents Reveal," National Security Archive Electronic Breifing Book, No. 195, July 31, 2006, http://www.gwu.edu/~nsarchiv/NSAEBB/NSA EBB195/index.htm, accessed Feb. 22, 2012.

11. *Radicals*

another activist: Interview, Matt Smucker, Dec. 23, 2012.

dual power: Interviews with Yotam Marom (Dec. 1, 2011), Eliot Tarver (Dec. 20, 2011), Meaghan Linick (Dec. 21, 2011).

in a hurry: See my *The Sixties,* p. 186.

"*a professional*": Yotam Marom at New School, Occupy Onwards conference, Dec. 18, 2011.

Citibank: Barbara Schneider Reilly, "Occupy Wall Street Protester, Arrested and Jailed for 30 Hours, Tells Her Story for the First Time," http://www.alternet.org/story/152882, accessed Dec. 29, 2011; Chris Bowers, "Occupy Wall Street: At least 19 arrested at Citibank branch," http://www.dailykos.com/story/2011/10/15/1026768/-Occupy-Wall-Street:-At-least-19-arrested-at-Citibank-branch, accessed Dec. 29, 2011.

New York magazine: John Heilemann, "2012 = 1968?" *New York,* Nov. 27, 2011, http://nymag.com/news/politics/occupy-wall-street-2011-12/index3.html, accessed Dec. 18, 2011.

OFS website: http://www.afreesociety.org/, accessed Dec. 27, 2011.

Hugo Chavez: "Hugo Chavez: Israel plans to 'terminate the Palestinian people,'" *Haaretz,* Nov. 28, 2009, http://www.haaretz.com/hasen/spages/1131227.html, and Meris Lutz, "Venezuela's Hugo Chavez slams Israel during Damascus visit," Sept. 4, 2009, http://latimesblogs.latimes.com/babylonbeyond/2009/09/syria-chavez-assad-slam-israel.html, accessed Dec. 28, 2011.

12. *The Co-optation Phobia*

often declared dead: This point was emphatically made by Frances Fox Piven and Dorian Warren at a forum on Occupy sponsored by *Dissent* and *Jacobin* magazines at Columbia University, Nov. 28, 2011.

"growing mobs": Amanda Terkel, "Eric Cantor Condemns Occupy Wall Street 'Mobs': They're 'Pitting Americans Against Americans,'" Nov. 7, 2011. http://www.huffingtonpost.com/2011/10/07/eric-cantor-occupy-wall-street-mobs_n_999853.html, accessed

Nov. 25, 2011. In subsequent days, however, Rep. Cantor declined the opportunity to repeat his charge. Apparently Republican pollsters thought he should mute the offense.

comic book Batman: Frank Miller, "Anarchy," Nov. 7, 2011, http://frankmillerink.com/2011/11/anarchy, accessed Nov. 25, 2011.

Ann Coulter: Leslie Rosenberg and Adam Shah, "Discussing Occupy Wall Street, Coulter Says: 'It Just Took A Few Shootings At Kent State To Shut That Down," Nov. 26, 2011, http://mediamatters.org/blog/201111260001, accessed Dec. 3, 2011.

"This dark carnival": E-mail from David Horowitz Freedom Center, Nov. 30, 2011.

"I love these people causin' trouble": http://watching-tv.ew.com/2011/10/25/david-letterman-occupy-wall-street-rush-limbaugh, accessed Dec. 2, 2011.

Elizabeth Warren: Beth Fuohy, "Occupy Wall Street And Democrats Remain Wary Of Each Other," *Huffington Post,* Nov. 17, 2011, http://www.huffingtonpost.com/2011/11/17/occupy-wall-street-democrats-2012-election_n_1099068.html, accessed Dec. 2, 2011.

Nancy Pelosi: David Weigel, "Pelosi Stands By Occupy Wall Street Without Naming It," *Slate.com,* Nov. 22, 2011, http://www.slate.com/blogs/weigel/2011/11/22/pelosi_stands_by_occupy_wall_street_without_naming_it.html, accessed Dec. 2, 2011.

she told Democrats: Brian Beutler, "Nancy Pelosi Games Out The Long Fight Over Medicare And The Rest Of The Safety Net," talkingpointsmemo.com, Dec. 7, 2011, http://tpmdc.talkingpointsmemo.com/2011/12/nancy-pelosi-games-out-the-long-fight-over-medicare-and-the-rest-of-the-safety-net.php?ref=fpne wsfeed, accessed Dec. 7, 2011. Apropos the Occupy movement, Pelosi had previously told Christiane Amanpour of ABC

News, "I support the message to the establishment, whether it's Wall Street or the political establishment and the rest, that change has to happen." Jessica Desvarieux, "Pelosi Supports Occupy Wall Street Movement," ABCnews.com, Oct. 9, 2011, http://abcnews.go.com/Politics/pelosi-supports-occupy-wall-street-movement/story?id=14696893#.Tt95t3NW6F4, accessed Dec. 7, 2011.

Justin Ruben: Telephone interview, Justin Ruben, Jan. 21, 2012.

"tiptoeing": Dennis Trainor, Jr., on http://groups.google.com/group/september17/browse_thread/thread/e6ccc21398a98d1a#, accessed Jan. 7, 2012.

"front group": Telephone interview, David DeGraw, Jan. 10, 2012.

harder-line movement's suspicion: Interview, Max Berger, Jan. 15, 2012.

Van Jones: Scott Wilson and Garance Franke-Ruta, "White House Adviser Van Jones Resigns Amid Controversy Over Past Activism," *Washington Post,* Sept. 6, 2009, http://voices.washingtonpost.com/44/2009/09/06/van_jones_resigns.html, accessed Dec. 31, 2011.

"what we're doing": Elliott, "Occupy Wall Street Looks Ahead."

"Everyone is jumping": John Heilemann, "2012 = 1968?" *New York,* Nov. 27, 2011, http://nymag.com/news/politics/occupy-wall-street-2011-12, accessed Dec. 31, 2011.

popular culture: The best discussion of this phenomenon is Joseph Heath and Andrew Potter, *Nation of Rebels: Why Counterculture Became Consumer Culture* (New York: HarperBusiness, 2004).

Priscilla Grim: "Occupy T-Shirts Disappear Amid Criticism," TMZ, Nov. 12, 2011, http://www.tmz.com/2011/11/12/jay-z-occupy-wall-street-t-shirts/#.TwsidCNSTgF, accessed Jan. 10, 2012.

the shirts were withdrawn: Elva Ramirez, "Jay-Z's 'Occupy All Streets' Shirt Vanishes from Rocawear Site Amid Controversy," Nov. 12, 2011, http://blogs.wsj.com/runway/2011/11/12/jay -zs-occupy-all-streets-shirt-vanishes-from-rocawear-site-amid -controversy/?mod=wsj_share_twitter, accessed Dec. 1, 2011.

"You have to engage": Interview, Jan. 8, 2012.

Miley Cyrus: http://www.youtube.com/watch?v=Ovs0fpF geqw, accessed Dec. 1, 2011. Her original music video of "Liberty Walk" is at jXRGpQH.

Tumblr website: http://wearethe99percent.tumblr.com, accessed Jan. 10, 2012.

told TMZ: "OCCUPY WALL STREET LEADER Calls BS on Miley Cyrus Vid —Singer is ALL TALK!," http:// www.tmz.com/2011/11/30/occupy-wall-street-miley-cyrus-vid eo-priscilla-grim/#.TthG83NW6F4, accessed Dec. 1, 2011.

Matt Smucker: Jonathan Matthew Smucker, "Radicals, Liberals & #OccupyWallStreet: This is What a Populist Alignment Looks Like," *BeyondtheChoir.org*, Oct. 12, 2011, http://beyond thechoir.org/diary/102/radicals-liberals-occupywallstreet-this-is -what-a-populist-alignment-looks-like, accessed Jan. 2, 2012.

told John Heilemann: Heilemann, "2012 = 1968?"

Part Three: The Promise

I asked Patrick: Interview, Jan. 3, 2011; and telephone interview, Jan. 7, 2011.

Ed Needham: Telephone interview, Jan. 5, 2011.

most of Europe: Jason DeParle, "Harder for Americans to Rise From Lower Rungs," *New York Times*, Jan. 4, 2012, http:// www.nytimes.com/2012/01/05/us/harder-for-americans-to-rise -from-lower-rungs.html, accessed Jan. 6, 2012.

Alan Greenspan: Greenspan, "International Financial Risk Management," remarks before the Council on Foreign Relations, Nov. 19, 2002, quoted in Phil Angelides et al., *Financial Crisis Inquiry Commission Report* (New York: PublicAffairs, 2011), p. 34.

"shocked": Greenspan, testimony before House Oversight Committee, Oct. 24, 2008, http://www.npr.org/templates/story/story.php?storyId=96070766, accessed Jan. 4, 2011.

more-equality: I borrow the phrase from Herbert J. Gans, *More Equality* (New York: Pantheon, 1973).

13. *Live-In Victories*

At Lincoln Center: James C. Taylor, "Composer Philip Glass joins Occupy Lincoln Center protest," *latimes.com,* Dec. 2, 2011, http://latimesblogs.latimes.com/culturemonster/2011/12/composer-philip-glass-joins-occupy-lincoln-center-protest.html, accessed Jan. 10, 2012.

"Occupy Octopus": "Rose Bowl parade gets occupied," http://www.cbsnews.com/8301-201_162-57350999/rose-bowl-parade-gets-occupied, accessed Jan. 6, 2012, and Sarah Armaghan, "Grand Central latest Occupy Wall Street destination; cops bust three in flash mob protesting defense act," *New York Daily News,* Jan. 3, 2012, http://www.nydailynews.com/new-york/grand-central-latest-occupy-wall-street-destination-cops-bust-flash-mob-protesting-defense-act-article-1.1 0005 84, accessed Jan. 6, 2012.

Bloomberg's townhouse: Nick Pinto, "Occupy Wall Street Goes to Bloomberg's House to Protest Press Arrests," http://blogs.villagevoice.com/runninscared/2012/01/occupy_wall_str_42.php, accessed Jan. 7, 2012.

homeless people in America: I paraphrase Rob Robinson,

co-founder of the Take Back the Land movement, interview, Dec. 20, 2011.

"It helps": Telephone interview, Jan. 11, 2012.

some will object: See, for example, critical comments posted at http://southcobb.patch.com/articles/occupy-cobb-to -protest-foreclosure-auctions-at-county-courthouse-today, culminating in Richard Pellegrino's Jan. 3, 2012, comment, accessed Jan. 12, 2012.

"driving force": Steven Pearlstein, " 'No Money Down' Falls Flat," *Washington Post,* March 14, 2007, http:// www.washingtonpost.com/wp-dyn/content/article/2007/03/13/ AR2007031301733_pf.html, accessed Jan. 12, 2012.

"firma, fecha": Edmund Conway, " 'Ninja' loans explode on sub-prime frontline," *Telegraph* (London), March 3, 2008, http://www.telegraph.co.uk/finance/economics/2785403/Ninja -loans-explode-on-sub-prime-frontline.html, accessed Jan. 10, 2012.

fake signatures: Pallavi Gogoi, "Robo-signing scandal may date back to late '90s," Associated Press, Sept. 1, 2011, http:// www.msnbc.msn.com/id/44365184/ns/business-real_estate/t/ robo-signing-scandal-may-date-back-late-s/#.Twy0xCNSTgE, accessed Jan. 10, 2011.

north Minneapolis: Jon Christian, "A House Divided: Occupy Homes Movement," campusprogress.org, Dec. 19, 2011, http://campusprogress.org/articles/a_house_divided_occupy_ho mes_movement, accessed Jan. 9, 2012.

Catherine Lennon: http://blip.tv/indy-tv/take-back-the-land -rochester-press-conference-speakout-and-vigil-on-3-28-11-49 64387, accessed Jan. 9, 2012.

won a court order: http://www.takebacktheland.org/index .php?mact=News,cntnt01,print,0&cntnt01articleid=7&cntnt 01showtemplate=false&cntnt01returnid=15, accessed Jan. 9, 2012.

"We're not slaves": Natasha Lennard, "Community Stands Strong to Block an Eviction," Aug. 19, 2011, http://city room.blogs.nytimes.com/2011/08/19/community-stands-strong -to-block-a-foreclosure, and http://www.o4onyc.org, accessed Jan. 9, 2012.

foreclosure auction: Ben Hallman and Michael Hudson, "'Occupy Wall Street' aims ire at foreclosures," Center for Public Integrity iwatch, Oct. 13, 2011,]http://www.iwatch news.org/2011/10/13/7104/occupy-wall-street-aims-ire-foreclo sures, accessed Jan. 9, 2012.

thirty-seven were arrested: Steven Thrasher, "Singing Protesters Arrested Again During 'Public' Foreclosure Auction," http:// blogs.villagevoice.com/runninscared/2012/01/singing_protest .php, accessed Feb. 1, 2012.

Brigitte Walker: http://www.youtube.com/watch?feature=play er_embedded&v=4GbnpTJoSWc, http://occupyatlanta.org/2011/ 12/19/interview-with-brigitte-walker-12-8-11-in-riverdale/#.Tw eFUiNSTgE and http://occupyourhomes.org/blog/2011/dec/ 20/brigitte-walker-victory, accessed Jan. 9, 2012.

"The Occupy kids": Interview, Rob Robinson, Dec. 20, 2011.

underwater: http://newspaper.occupybk.org/2011/12/16/up dates-from-702-vermont-street,http://www.homes.com/Home -Prices/ID-400019718825/702-VERMONT-ST, accessed Jan. 8, 2012.

they had torn down: Candice M. Giove, "'They took my place!' Single dad trying to take back home occupied by OWS," *New York Post,* Jan. 15, 2012, http://www.nypost.com/p/news/ local/brooklyn/ows_home_invasion_z9ApqDP6Q0boFviq8Cjv AL#ixzz1l8LuplTi, accessed Feb. 1, 2012; e-mail from Karanja Gaçuça, Feb. 1, 2012.

Boise, Idaho: Katy Moeller, "Occupy Wall Street groups in Boise, elsewhere target foreclosed properties," *Idaho Statesman,* Dec. 7, 2011, http://www.idahostatesman.com/2011/

12/07/1907250/occupy-wall-street-groups-target.html, accessed Jan. 12, 2012. Another example of anti-foreclosure action, this one in Southern California, is described in "Movement changes shape on shifting political sands," *Sydney Morning Herald,* Jan. 7, 2012, http://www.smh.com.au/world/movement-changes-shape -on-shifting-political-sands-20120106-1pobz.html, accessed Jan. 10, 2012.

Occupy Oakland: John C. Osborn, "Police disperse West Oakland Occupy site," Dec. 28, 2011, http://oaklandnorth .net/2011/12/28/police-disperse-west-oakland-occupy-site, ac-cessed Jan. 12, 2012.

14. *Work the System? Change It? Smash It?*

underdeveloped: More generally, I believe with the sociolo-gist Duncan Watts (*Everything Is Obvious: *Once You Know the Answer* [New York: Crown Business, 2011]) that the time has long since arrived for the pretensions of social science to fade away gracefully. But this is ground to be plowed elsewhere.

Karl Rove: Crossroads GPS, " 'Foundation' Ma," Nov. 9, 2011, http://www.youtube.com/watch?feature=player_embedded&v =tNxez4ddpa0, accessed Jan. 10, 2012.

Warren replied: Chris Bowers, "Elizabeth Warren re-sponds to Karl Rove, stands by Occupy Wall Street," DailyKos, Nov. 10, 2011, http://www.dailykos.com/story/20 11/11/10/1035145/-Elizabeth-Warren-responds-to-Karl-Rove, -stands-by-Occupy-Wall-Street, accessed Jan. 10, 2012.

Frank Luntz: Chris Moody, "How Republicans are being taught to talk about Occupy Wall Street," Yahoo News, Dec. 1, 2011, http://news.yahoo.com/blogs/ticket/republicans -being-taught-talk-occupy-wall-street-133707949.html, ac-cessed Jan. 10,2012.

"Labor Day": Darlene Superville, "Obama Labor Day Speech: President Says Congress Must Pass Jobs Plan," *Huffington Post,* Sept. 5, 2011, http://www.huffingtonpost.com/2011/09/05/obama-labor-day-speech-_n_949374.html, accessed Dec. 31, 2011.

William Daley: David Weigel, "Bill Daley on Business and #OccupyWallStreet," *Slate,* Oct. 5, 2012, http://www.slate.com/blogs/weigel/2011/10/05/bill_daley_confidence_man.html, accessed Jan. 11, 2012.

"tiptoeing": Benjy Sarlin, "Obama: Occupy all Street Is 'Giving Voice To A More Broad Based Frustration,'" *Talking Points Memo,* Oct. 6, 2011, http://2012.talkingpointsmemo.com/2011/10/obama-occupy-wall-street-is-giving-voice-to-a-more-broad-based-frustration.php, accessed Dec. 31, 2011.

by November 22: David Edwards, "Obama gets 'mic checked' in New Hampshire, Nov. 22, 2011, http://www.rawstory.com/rs/2011/11/22/obama-gets-mic-checked-in-new-hampshire, accessed Jan. 9, 2012.

He adapted: Barack Obama, Keynote Speech to the 2004 Democratic Convention, http://www.usconstitution.net/obama.html, accessed Dec. 7, 2011.

"I believe that": Remarks by the President on the Economy in Osawatomie, Kansas, Dec. 6, 2011, http://www.whitehouse.gov/the-press-office/2011/12/06/remarks-president-economy-osawatomie-kansas, accessed Dec. 7, 2011.

Richard J. Daley: I've written at length about the Chicago clashes in *The Whole World Is Watching: Mass Media in the Making and Unmaking of the New Left* (Berkeley: University of California Press, 1980), especially pp. 186-189, and *The Sixties: Years of Hope, Days of Rage* (New York: Bantam, 1987), especially pp. 319-36.

Don Rose: Don Rose, "Emanuel's Incipient, Self-Engineered Train Wreck," *Chicago Daily Observer,* Jan. 10, 2012, http://

www.cdobs.com/archive/featured/emanuels-incipient-self-en gineered-train-wreck, accessed Jan. 11, 2012.

Chicago aldermen: Telephone interview, Don Rose, Jan. 14, 2012.

Meaghan Linick: Interview, Dec. 21, 2011.

French tract: http://libcom.org/files/thecominsur_booklet [1].pdf, accessed Jan. 13, 2012.

Tampa's: Rania Khalek, "Tanks, SWAT Teams, Surveillance Helicopters: Cities Already Turning Into Mini-Police States for the Political Conventions," *Alternet,* Jan. 10, 2012, http://www .alternet.org/story/153730/tanks%2C_swat_teams%2C_surveil lance_helicopters%3A_cities_already_turning_into_mini-police _states_for_the_political_conventions?page=entire, accessed Jan. 12, 2012.

Charlotte banned: Steve Harrison, "Charlotte's Democratic National Convention ordinances worry the ACLU," *Charlotte Observer,* Jan. 9, 2012, http://www.charlotteobserver .com/2012/01/09/2910033/charlotte-dnc-ordinances-aclu.html, accessed Jan. 12, 2012.

Cage peaceful: Two recent books underscore how "the right of the people peaceably to assemble" has been rolled back over the last four decades. For a summary, see Jeremy Kessler, "The Closing of the Public Square," *The New Republic,* "The Book," http://www.tnr.com/book/review/the-closing-the-public -square-john-inazu-timothy-zick, accessed Jan. 12, 2012.

On November 8: Alliance of Community Trainers (Starhawk, Lisa Fithian, Lauren Ross [or Jupiter]), "Open Letter to the Occupy Movement: Why We Need Agreements," Nov. 8, 2011, http://trainersalliance.org/?p=221, accessed Jan. 13, 2012.

Bill Dobbs: Justin Elliott, "Occupy Wall Street looks ahead to 2012," *Salon,* Nov. 11, 2011, http://www.salon.com/2011/11/11/ occupy_wall_street_looks_ahead_to_2012/singleton/, accessed Jan. 12, 2012.

"occupy the agenda": Sarah Seltzer, "Occupiers aren't running for office. They have their sights set higher," *Washington Post,* Jan. 13, 2012, http://wapo.st/xkpoM7, accessed Jan. 14, 2012.

One candidate: Dylan Byers, "The first Occupy candidate: Nate Kleinman," *Politico,* Jan. 24, 2012, http://ww.politico .com/blogs/media/2012/01/the-first-occupy-candidate-nate-klein man-112057.html, accessed Feb. 1, 2012.

15. *Can the Outer Movement Get Organized?*

numbers to support them: Telephone interview, Adam Green, Progressive Change Campaign Committee (PCCC), Jan. 19, 2012. PCCC claims 850,000 members.

Marty Frates: Telephone interview by Travis Mushett, Jan. 12, 2012.

Working America: http://ninedemands.com/petitions/work ing-america, accessed Jan. 13, 2012.

6.9 percent: Bureau of Labor Statistics press release, Jan. 21, 2011, http://www.bls.gov/news.release/union2.nr0.htm, accessed Jan. 13, 2012.

Transport Workers: Jen Doll, "Bus Drivers Should Not Have to Transport Arrested Occupy Wall Street Protesters, Says TWU," http://blogs.villagevoice.com/runninscared/2011/10/ occupy_wall_str_10.php, accessed Feb. 22, 2012.

On the left coast: Laura Clawson, "Occupy Oakland attempting port shutdown despite union opposition," Daily Kos, Dec. 12, 2012, http://www.dailykos.com/story/2011/12/12/1044458/-Oc cupy-Oakland-attempting-port-shutdown-despite-union-opposi tion; Elizabeth Flock, "Occupy Oakland plans West Coast port shutdown, but port workers don't support it," *Washington Post* online, Dec. 5, 2011, http://www.washingtonpost.com/blogs/

blogpost/post/occupy-oakland-plans-west-coast-port-shutdown
-but-port-workers-dont-support-it/2011/ 12/05/gIQAJLEbWO_
blog.html-americas-truck-drivers-occupy-ports, accessed Jan. 14,
2012.

A group of them: Leonardo Mejia et al., "An Open Letter
from America's Truck Drivers on Occupy the Ports," Dec. 13,
2011, http://westcoastportshutdown.org/content/open-letter
-americas-truck -drivers-occupy-ports, accessed Jan. 14, 2012.

"When all": Telephone interview, Dec. 26, 2011.

IRS regulations: Internal Revenue Service, "Employee vs.
Independent Contractor—Seven Tips for Business Owners,"
http://www.irs.gov/newsroom/article/0,,id=173423,00.html, accessed Jan. 14, 2012.

An executive: Alison Vekshin, James Nash and Susanna Ray,
"Goldman Sachs Target as Occupy Protesters Curb Some West
Coast Shipping," Bloomberg News, Dec. 13, 2011, http://www
.bloomberg.com/news/2011-12-12/goldman-sachs-top-target-of
-occupy-protests-at-west-coast-ports.html, accessed Jan. 14,
2012.

16. *Is There a Global Revolution?*

Anthony Barnett writes, "original": Anthony Barnett, "The
Long and the Quick of Revolution," *openDemocracy,* Dec. 16,
2011, http://www.opendemocracy.net/anthony-barnett/long
-and-quick-of-revolution, accessed Jan. 15, 2012.

Bouazizi: http://www.youtube.com/watch?v=sTah7XWyMDE
and http://nos.nl/video/209867-begrafenis-mohammed-bouazizi
.html, accessed Jan. 15, 2012.

Facebook page: http://www.facebook.com/pages/Khaled
-Said/100792786638349?sk=photos, accessed Jan. 15, 2012.

Tom Nairn's phrase: Tom Nairn, "Hooligans of the Abso-

lute: Black Pluto's door after 11 September," *openDemocracy,*
Oct. 3, 2001, http://www.opendemocracy.net/democracy-glo
baljustice/article_155.jsp, accessed Jan. 15, 2012.

17. *"This Is the Beginning of the Beginning"*

unfairly favors the wealthy: Pew Research Center for the
People and the Press, "77% - Public Views of Inequality, Fairness and Wall Street," Jan. 6, 2012, http://pewresearch.org/
databank/dailynumber/?NumberID=1400, accessed Jan. 11,
2012.

Nona Willis Aronowitz: "In an OWS Era, Americans Are
Much More Aware of Class Tension," *Good News,* Jan. 12,
2012, http://www.good.is/post/in-an-ows-era-americans-are-a
-lot-more-aware-of-class-tension, accessed Jan. 15, 2012.

Frances Fox Piven: Symposium on "Occupy Wall Street—
Phase II," December 5, 2011, http://dissentmagazine.org/atw
.php?id=629, accessed Jan. 16, 2012.

Afterword

fifteen or twenty thousand: Noam Chomsky, "Tom-
gram: Noam Chomsky, A Rebellious World or a New Dark
Age," May 8, 2012, www.tomdispatch.com/post/175539/
tomgram%3A_noam_chomsky%2C_a_rebellious_world_or_a_
new_dark_age, accessed May 8, 2012.

In the Bay Area: "May Day Protests Around the Bay,"
May 1, 2012, blogs.kqed.org/newsfix/2012/05/01/bay-area
-may-day-protests, accessed May 8, 2012.

Taken aback: Liz Melchor, "OccupySF Condemns Mission
Vandalism," May 7, 2012, missionlocal.org/2012/05/occupysf

-reacts-to-monday-nights-destruction-of-valencia, accessed May 8, 2012.

"we were hijacked": Scott Rossi, "Notes from an Occupation 17: Dolores Park 'Ruckus,'" May 1, 2012, scottrossi.tumblr .com/post/22184158717/notes-from-an-occupation-17-dolores -park-ruckus, accessed May 8, 2012.

decades ago: The Whole World Is Watching: Mass Media in the Making and Unmaking of the New Left (University of California Press, 1980), pp. 28, 271.

on Facebook: For example, www.facebook.com/photo.php ?fbid=3621355328590&set=p.3621355328590&type=1&theater and www.facebook.com/photo.php?fbid=3637163963796&set= a.1203664207823.2032934.1115158246&type=1&theater, accessed May 8, 2012.

photos showed: For example, www.nytimes.com/slideshow/ 2012/05/04/nyregion/20120504wip-6.html, accessed May 9, 2012.

New York Daily News: Rocco Parascandola and Shayna Jacobs, "NYPD arrests 86 in Occupy Wall Street protests," May 2, 2012, www.nydailynews.com/new-york/nypd-arrests-86 -occupy-wall-street-protests-article-1.1071557, accessed May 9, 2012.

New York Times *online headline:* Colin Moynihan, *New York Times* City Room blog, May 2, 2012, cityroom.blogs.nytimes .com/2012/05/02/scores-cuffed-or-cited-by-end-of-day-of-demon strations, accessed May 9, 2012. Moynihan is a good reporter, and it deserves repeating that reporters don't write their own headlines.

Occupy Wall Street largely failed: "Goodbye, Occupy," *New York Post,* May 2, 2012, www.nypost.com/p/news/opinion/ editorials/goodbye_occupy_U6Xgt1BI2fUN4uh2nJcnDO, accessed May 17, 2012.

One shareholder: Jennifer Carillo, quoted in Ryan Pitkin,

"They were loud, they were proud, they were the 99 percent," *Creative Loafing Charlotte,* clclt.com/charlotte/they-were-loud-they-were-proud-they-were-the-99-percent/Content?oid=2713320, accessed May 10, 2012.

92 percent: Kirsten Valle Pittman and Andrew Dunn, "Bank of America CEO defends bank's leadership," *Charlotte Observer,* May 9, 2012, www.charlotteobserver.com/2012/05/09/3229324/bank-of-america-meeting-draws.html, accessed May 10, 2012.

a different approach: I sketched an earlier version of these concluding remarks on the Brainstorms blog of the *Chronicle of Higher Education* in "Is Freedom of Assembly a Dead Letter?" May 7, 2012, chronicle.com/blogs/brainstorm/is-freedom-of-assembly-a-dead-letter, accessed May 10, 2012. The tenor of the comments posted there illustrates how contested—and important—the freedom of assembly is.

James Madison: Speech in Congress on "Self-Created Societies," November 27, 1794, www.constitution.org/jm/1794 1127_societies.htm, accessed May 10, 2012.

More evidence: Jim Gold, "Mayors deny colluding on 'Occupy' crackdowns," November 11, 2011, www.msnbc.msn.com/id/45312298/ns/us_news-life/t/mayors-deny-colluding-occupy-crackdowns/#.T6foF59YtDI, accessed May 10, 2012.

John D. Inazu: New Haven: Yale University Press, 2012.

Inazu's book: Zick's book was published by Cambridge University Press in 2008. The Inazu and Zick books are ably reviewed by Jeremy Kessler in *The New Republic,* "The Closing of the Public Square," January 12, 2012, www.tnr.com/book/review/the-closing-the-public-square-john-inazu-timothy-zick, accessed May 10, 2012.